THE STORY OF NASA'S CREW 8

Collaboration Between Space Agencies and Private Enterprises

MARCUS T. HOOKS

COPYRIGHT

Copyright©2024 Marcus T. Hooks. All rights reserved. No part of this publication may be reproduced, distributed, or transmitted in any form or by any means, including photocopying, recording, or other electronic or mechanical methods, without the prior written permission of the publisher, except in the case of brief quotations embodied in critical reviews and certain other non-commercial uses permitted by copyright law

TABLE OF CONTENTS

COPYRIGHT 1
TABLE OF CONTENTS 2
INTRODUCTION 4
    The Story of NASA's Crew 8 4
CHAPTER 1 11
    Introduction to the Crew-8 Mission 11
CHAPTER 2 20
    The Importance of International Space Missions 20
CHAPTER 3 31
    Life on the ISS 31
CHAPTER 4 42
    The Science Behind Weather Delays 42
CHAPTER 5 53
    Hurricane Milton and its Impact 53
CHAPTER 6 73
    Crew-8: The Astronauts and Cosmonaut 73
CHAPTER 7 83
    SpaceX's Role in Human Spaceflight 83
CHAPTER 8 95
    The Return to Earth: Challenges and Triumphs 95
CHAPTER 9 106
    Future of Space Travel: What's Next After Crew-8? .. 106

CONCLUSION 116
    NASA'S CREW 8 116

INTRODUCTION
The Story of NASA's Crew 8

Space exploration has always been a story of courage, technology, and international cooperation. At the heart of these missions lies not only the quest for scientific discovery but also the unyielding determination of astronauts and engineers. *The Story of NASA's Crew 8* is a reflection of these themes, encapsulating the challenges and triumphs of one of the most recent International Space Station (ISS) missions.

The tale of Crew-8 is an embodiment of resilience—an account of a team that weathered not just the challenges of life in space but also the unpredictable forces of nature. As with many space missions, the success of Crew-8 did not come without obstacles. Their journey began as part of a joint operation between NASA and SpaceX, marking another milestone in the partnership between governmental space agencies and private enterprise. The team's mission was to continue groundbreaking scientific research aboard the ISS, fostering advancements in various fields such as biology, physics, and earth sciences. However, the final chapter of their expedition was far from routine.

After successfully completing months of experiments and duties aboard the ISS, Crew-8's return to Earth was hindered by a series of unforeseen weather delays. Nature's interference, in the form of Hurricane Milton, created hazardous conditions that forced NASA and SpaceX to delay the scheduled splashdown. Such delays underscore the complexity of modern space travel, where even the most advanced technologies are subject to the whims of the natural world. This book offers a deep dive into not just the mission's technical aspects but also the emotional and psychological toll on the astronauts as they prepared for an extended stay in space beyond their planned return.

The Crew: A Glimpse into Human Endurance

At the core of *The Story of NASA's Crew 8* are the astronauts themselves—individuals who have dedicated their lives to advancing human knowledge through space exploration. The Crew-8 team included a mix of NASA astronauts and a Russian cosmonaut, reflecting the ongoing international collaboration that defines space missions today. These men and women were more than just passengers on a spacecraft; they were the scientists, engineers, and explorers tasked with carrying out experiments that could impact life on Earth for decades to come.

Among the key figures in Crew-8 were NASA's Michael Barratt, Jeanette Epps, and Matthew Dominick, alongside Russian cosmonaut Alexander Grebenkin. Each of these astronauts brought a unique set of skills and experiences to the mission, which proved essential during their extended stay aboard the ISS. Throughout their mission, these individuals conducted experiments ranging from medical research aimed at understanding how space affects the human body to technological innovations that could improve satellite communications and climate monitoring.

While the astronauts were no strangers to rigorous schedules and isolation, the delay in their return added an unexpected layer of endurance. The psychological impact of prolonged missions, coupled with the uncertainty of their return date, tested the crew's mental resilience. The extended stay required not just physical stamina but also emotional flexibility, as they navigated the complexities of space travel while staying in constant communication with their families and mission control. These human elements—often overshadowed by the technicalities of spaceflight—are central to the Crew-8 story.

SpaceX and NASA: A Collaboration in Innovation

The collaboration between NASA and SpaceX on Crew-8 marks a new era in space exploration. SpaceX, the private aerospace company founded by Elon Musk, has been instrumental in reigniting public interest in space travel with its innovative reusable rockets and spacecraft designs. For Crew-8, SpaceX's Crew Dragon capsule, known as "Endeavour," played a critical role. This spacecraft represented the latest in human spaceflight technology, capable of docking autonomously with the ISS and carrying crew members safely to and from space.

NASA, with its decades of experience in space missions, brought to the table its unparalleled expertise in managing human spaceflight programs. This partnership highlights a new model for space exploration—one that blends the capabilities of public and private sectors to push the boundaries of what humanity can achieve in space.

Crew-8's mission was emblematic of this new collaboration. Launched on a Falcon 9 rocket, the crew journeyed to the ISS to carry out their scientific objectives. However, the return journey—marked by numerous delays—became a testament to the complex coordination between NASA, SpaceX, and weather-monitoring agencies. As the crew awaited better conditions, mission planners worked tirelessly to ensure the astronauts' safe return, demonstrating the immense logistics involved in modern space travel.

The Science of Weather and Space Travel

While we often think of space exploration as being far removed from Earth's elements, the reality is that space missions are deeply affected by our planet's weather systems. The safe return of astronauts from the ISS requires precise timing, particularly when it comes to splashdown zones in the ocean. Crew-8's delayed return was due in large part to Hurricane Milton, which churned through the Atlantic and Gulf of Mexico during the crew's scheduled departure.

This aspect of the story shines a light on the interconnectedness of Earth's climate systems and space travel. Despite all the advancements in technology, weather remains a dominant factor in determining when and how astronauts can safely return to Earth. The importance of weather forecasting in space missions cannot be overstated. For Crew-8, the delay due to rough seas and dangerous winds illustrates the delicate balance between human ambition and natural forces.

In the context of space exploration, hurricanes and storms present unique challenges. While modern spacecraft are built to withstand the rigors of space travel, re-entering Earth's atmosphere and landing safely requires calm seas and favorable wind conditions. This book delves into how meteorologists, flight engineers, and mission planners work together to monitor weather patterns, adjust timelines, and ensure that astronauts are not placed at undue risk during their re-entry.

The Future of Space Travel: Lessons from Crew-8

The story of Crew-8 offers a glimpse into the future of space exploration. As NASA and SpaceX continue to push the boundaries of human spaceflight, lessons from this mission will inform future expeditions. Crew-8 highlighted the importance of collaboration between public and private entities in space travel, demonstrating that the future of exploration is built on partnerships.

Moreover, the challenges faced by Crew-8 underscore the need for continuous improvement in spacecraft technology, weather forecasting, and mission planning. Space travel is not without its risks, but each mission, successful or delayed, brings new insights that move humanity closer to the next frontier—whether that be a return to the Moon or a crewed mission to Mars.

Crew-8's extended stay aboard the ISS also offers valuable lessons in human resilience. Space is an unforgiving environment, and the psychological challenges of isolation, uncertainty, and distance from loved ones are just as significant as the technical hurdles of space travel. As space missions grow longer and more complex, understanding how to support astronauts' mental health will be critical for future explorations.

A Mission Worth Remembering

In *The Story of NASA's Crew 8,* we explore not only the technical and scientific aspects of the mission but also the human element. From the astronauts who braved months in space to the mission control teams who navigated the complexities of their return, every part of the Crew-8 story is a testament to human determination, collaboration, and ingenuity. The mission may have been delayed, but the spirit of exploration and discovery was never dampened.

As we look to the future of space travel, the lessons learned from Crew-8 will pave the way for the next generation of astronauts, scientists, and engineers who will take humanity even further into the cosmos.

CHAPTER 1
Introduction to the Crew-8 Mission

The Crew-8 mission represents another significant step in the evolution of human spaceflight, highlighting NASA's continued collaboration with private space company SpaceX. This partnership, which began in 2014 through NASA's Commercial Crew Program, marks a shift in how space missions are conducted, leveraging the strengths of both government agencies and private companies to achieve ambitious goals in space exploration.

SpaceX, founded by Elon Musk, has become a major player in the space industry, with its reusable rocket technology and its Dragon spacecraft at the forefront of modern space missions. NASA, with its rich legacy of human spaceflight, saw the opportunity to tap into the innovation of private companies like SpaceX to continue its goal of advancing space science, technology, and exploration. The Crew-8 mission, which was part of this collaboration, is a perfect example of how public-private partnerships can work to push the boundaries of what's possible in space.

NASA's Goals for the Crew-8 Mission

The Crew-8 mission, like many of NASA's recent ventures, was centered around both scientific research and international collaboration. The International Space Station (ISS) remains one of the most important assets in the study of microgravity, Earth sciences, and human biology in space. With Crew-8, NASA aimed to continue and expand its research on these fronts. The astronauts on board conducted numerous scientific experiments that are intended to further our understanding of space and its effects on human health, technology, and biology.

One of the key goals of Crew-8 was to carry out long-term research to benefit life on Earth. The experiments covered various fields, including biological sciences, environmental monitoring, and physical sciences. Research into how living organisms react to space conditions, such as plants and human cells, is especially important for future missions to the Moon and Mars, where humans will need to live and work for extended periods. The knowledge gained from the Crew-8 mission will contribute to building a strong foundation for these future exploratory missions.

Beyond scientific exploration, NASA also focused on maintaining the ISS's functionality and performing upgrades to its systems. Routine maintenance and upgrades are critical to ensuring that the ISS continues to be a viable platform for scientific research and international cooperation for many years to come. Crew-8 played a vital role in maintaining this continuity by supporting the long-term viability of the ISS through technical work and repairs.

SpaceX's Role in the Crew-8 Mission

SpaceX's role in the Crew-8 mission was pivotal, as it provided the transportation for both astronauts and essential cargo. The Crew Dragon spacecraft, known as *Endeavour*, was the vehicle that took the crew to the ISS and brought them back to Earth. Launched atop a Falcon 9 rocket, the Crew Dragon demonstrated the effectiveness and safety of reusable space vehicles, continuing SpaceX's tradition of pushing innovation in space travel.

The ability of the Crew Dragon to autonomously dock with the ISS is a technological leap that has helped minimize the need for manual intervention, providing a safer and more efficient way to transport astronauts. The collaboration between NASA and SpaceX in the Crew-8 mission showcases the future of space exploration, where the strengths of public and private sectors are merged to achieve common goals.

SpaceX's involvement in human spaceflight through the Crew Dragon missions signifies a critical shift in how we approach space exploration. The use of reusable rockets and spacecraft reduces costs significantly and improves the frequency of missions, allowing for more frequent trips to space. This approach represents a fundamental shift in space travel, where the economics of space exploration become more sustainable over the long term.

The Astronauts of Crew-8

The Crew-8 mission brought together a diverse team of astronauts from different backgrounds, each contributing their expertise to the success of the mission. The crew comprised NASA astronauts Michael Barratt, Matthew Dominick, and Jeanette Epps, along with Roscosmos cosmonaut Alexander Grebenkin. This team not only highlighted the international cooperation essential to space missions but also demonstrated the deep collaboration between countries like the United States and Russia, even in times of geopolitical tension.

Michael Barratt

Michael Barratt, the most experienced of the crew, served as a key figure on the mission. A veteran astronaut, Barratt has spent hundreds of days in space over multiple missions, contributing significantly to scientific research on human spaceflight. His background as a physician made him particularly valuable in conducting medical and biological experiments aboard the ISS.

Jeanette Epps

Jeanette Epps, an aerospace engineer and mission specialist, brought a wealth of technical expertise to the mission. Epps' role in Crew-8 represented her first spaceflight, and she contributed significantly to various technical projects and research studies during the mission. Her participation also marked an important milestone in NASA's effort to promote diversity and inclusion in space exploration.

Matthew Dominick

Matthew Dominick, another key member of the crew, played a pivotal role in conducting the day-to-day operations aboard the ISS. His background as a test pilot provided valuable skills in managing the Crew Dragon spacecraft's systems and ensuring the safety of the mission.

Alexander Grebenkin

The inclusion of Russian cosmonaut Alexander Grebenkin was a continuation of the long-standing collaboration between NASA and Roscosmos, Russia's space agency.

Grebenkin's presence underscored the international nature of the ISS and space exploration as a whole. His contributions were instrumental in various technical tasks aboard the station and in maintaining the operational efficiency of the ISS.

The Mission's Timeline and Achievements

Crew-8 was launched on March 3, 2024, from NASA's Kennedy Space Center aboard a Falcon 9 rocket. The mission began with a flawless ascent into space, followed by a smooth docking of the Crew Dragon spacecraft with the ISS two days later. The crew spent several months aboard the ISS, conducting scientific experiments, performing maintenance tasks, and engaging in outreach activities with students and the public.

One of the mission's major achievements was its contribution to Earth sciences. Crew-8's experiments included monitoring changes in the Earth's atmosphere, studying climate patterns, and observing environmental shifts from space. These findings are vital for understanding the long-term effects of climate change and for improving our ability to monitor and protect the planet.

In addition to Earth sciences, the Crew-8 mission also focused on medical research. Experiments on human health and physiology in microgravity are essential for preparing astronauts for longer missions, such as those to Mars. The results of these studies provide invaluable data on how the human body responds to extended periods of time in space, which is critical for developing countermeasures to keep astronauts healthy during long-term missions.

Another notable achievement of Crew-8 was its support for technology development. The crew tested new technologies that will be crucial for future space exploration missions, including advanced communications systems, tools for maintaining spacecraft, and systems for growing food in space. These experiments contribute to building the infrastructure needed for sustained human presence in space.

Overcoming Challenges: The Weather Delays

One of the most significant aspects of the Crew-8 mission was the weather delays that affected the crew's return to Earth. Originally scheduled to depart from the ISS in early October 2024, the crew's return was delayed multiple times due to unfavorable weather conditions in the planned splashdown area off the coast of Florida. Hurricane Milton, in particular, presented major challenges, as it caused dangerous seas and winds that made a safe splashdown impossible.

This delay tested the resilience of both the astronauts and the mission planners at NASA and SpaceX. The extended stay aboard the ISS required careful management of resources, including food, water, and air, and highlighted the unpredictability of space missions. The crew's safe return, despite the challenges, showcased the meticulous planning and adaptability of everyone involved in the mission.

A New Chapter in Space Exploration

The Crew-8 mission was not just a routine trip to the ISS—it represented the next step in the evolution of human spaceflight. Through collaboration between NASA and SpaceX, it demonstrated that public-private partnerships could achieve extraordinary results. The mission's scientific achievements, its contribution to international cooperation, and the resilience of its crew in the face of delays make it a story worth telling.

As NASA and SpaceX continue to push the boundaries of space exploration, missions like Crew-8 provide a blueprint for the future. The lessons learned from this mission will inform future expeditions to the Moon, Mars, and beyond, ensuring that humanity's reach into space continues to grow.

The Crew-8 mission not only fulfilled its scientific and exploratory goals but also strengthened the relationship between human spaceflight and technological innovation. It stands as a testament to the perseverance and ingenuity of the astronauts, scientists, and engineers who continue to drive the future of space exploration forward.

CHAPTER 2

The Importance of International Space Missions

Space exploration has always been an endeavor that transcends national boundaries, requiring the collaboration of many countries and their space agencies to achieve ambitious goals. The success of international space missions, particularly those involving long-term human habitation aboard the International Space Station (ISS), underscores the importance of global cooperation in advancing our understanding of space and enabling future exploration beyond Earth's orbit. This chapter will explore the significance of these collaborations, focusing on the partnerships between NASA and Roscosmos, as well as the broader context of international cooperation in space.

A History of Collaboration: NASA and Roscosmos

The partnership between NASA and Roscosmos, the Russian space agency, dates back to the early days of space exploration, when geopolitical rivalries between the United States and the Soviet Union gave rise to the Space Race. This competitive atmosphere culminated in landmark achievements such as the Soviet Union's launch of the first human, Yuri Gagarin, into space in 1961, and NASA's Apollo 11 mission that landed humans on the Moon in 1969. While these early years were marked by rivalry, they also laid the groundwork for future cooperation.

In the aftermath of the Cold War, both countries recognized the benefits of working together to share knowledge, resources, and expertise in space exploration. The construction of the ISS, which began in 1998, is one of the most visible and successful examples of this collaboration. Over two decades later, the ISS remains a symbol of international unity, hosting astronauts from multiple countries who work together on scientific research, technological development, and space exploration initiatives.

NASA and Roscosmos' partnership is foundational to the ISS's success. Russian Soyuz spacecraft played a critical role in ferrying astronauts to and from the ISS for years, particularly after the retirement of NASA's Space Shuttle program in 2011. The ISS's modular design, with its various components built and launched by different countries, reflects a truly global effort. NASA and Roscosmos have shared responsibilities for maintaining the station, conducting spacewalks, and managing scientific experiments, exemplifying the benefits of international cooperation.

Why International Cooperation Matters

Space exploration is an inherently expensive and resource-intensive endeavor. The costs of developing spacecraft, launching missions, and supporting astronauts in space are immense. By collaborating, space agencies can pool their resources, share costs, and distribute the workload, making ambitious projects like the ISS feasible. For example, the ISS is not solely a NASA or Roscosmos endeavor; it involves contributions from other space agencies, including the European Space Agency (ESA), Japan's JAXA, and the Canadian Space Agency (CSA). Each country brings its strengths to the table, whether it's technological innovation, funding, or specialized expertise.

This collaborative approach also allows for a greater exchange of scientific knowledge. The research conducted on the ISS spans a wide range of disciplines, from human biology to climate science. By involving multiple countries, the scope of the research can be expanded, and the results can benefit a larger global community. Research conducted on the ISS often has implications beyond space exploration, contributing to advancements in medicine, environmental monitoring, and technology that directly benefit people on Earth.

Moreover, international space missions help foster diplomatic relations. Despite political tensions between countries, space agencies have often maintained strong partnerships. The collaboration between NASA and Roscosmos, for example, has continued even during periods of strained political relations between the United States and Russia. The ISS serves as neutral ground where science and exploration take precedence over geopolitical concerns. This spirit of cooperation has the potential to build trust and open new channels of communication between nations.

The Role of Roscosmos in Space Exploration

Roscosmos has been an essential player in human spaceflight, both before and after the dissolution of the Soviet Union. Its Soyuz spacecraft, which has been in service since the 1960s, is one of the most reliable and time-tested human spaceflight vehicles in the world. After NASA retired its Space Shuttle fleet in 2011, the Soyuz became the sole method of transporting astronauts to the ISS until SpaceX's Crew Dragon became operational in 2020. Roscosmos' experience in space station operations, having previously managed the Soviet-era Mir Space Station, has been invaluable in maintaining and expanding the ISS.

Russian cosmonauts and engineers bring unique expertise to the ISS missions. For example, Roscosmos is responsible for operating the Zvezda Service Module, which provides life support systems such as oxygen generation, water supply, and temperature regulation. Roscosmos' role in spacewalks, cargo resupply missions, and technical maintenance has been instrumental in ensuring the ISS's continuous operation over the years.

Russia's contributions to space exploration extend beyond human spaceflight. The country has a long history of robotic exploration missions to the Moon, Mars, and Venus, and it continues to play a role in global scientific efforts to explore the solar system. Roscosmos' ongoing missions, including collaboration with NASA on lunar exploration through the Artemis program, underscore its critical role in advancing space science.

The Future of NASA and Roscosmos Collaboration

As space exploration continues to evolve, NASA and Roscosmos will likely remain key partners, though the nature of their collaboration may shift as new players and technologies emerge. The rise of private space companies, particularly in the United States, is changing the landscape of space exploration. Companies like SpaceX and Blue Origin are now capable of launching astronauts and cargo to the ISS, reducing NASA's reliance on Roscosmos for transportation.

However, Roscosmos remains an essential partner for long-duration space missions, particularly as the world looks toward more ambitious goals, such as returning humans to the Moon and sending crewed missions to Mars. NASA's Artemis program, which aims to land the next astronauts on the Moon by the mid-2020s, includes potential collaboration with Roscosmos through the Lunar Gateway, a space station that will orbit the Moon and serve as a staging point for missions to the lunar surface and beyond. This partnership would combine Russia's experience with space station operations and NASA's advancements in crewed spaceflight and lunar exploration.

As international partnerships expand to include more countries and private companies, NASA and Roscosmos will continue to play leadership roles. The spirit of collaboration will remain central to their efforts, whether it involves maintaining the ISS, exploring new celestial bodies, or ensuring that humanity's presence in space continues to grow.

Expanding Global Cooperation: New Space Nations

While NASA and Roscosmos are two of the largest and most established space agencies, many other countries are making significant contributions to space exploration. The European Space Agency (ESA), Japan's JAXA, and the Canadian Space Agency (CSA) are longtime partners in human spaceflight and robotic exploration. However, new players are also emerging on the global stage. Countries like China, India, and the United Arab Emirates (UAE) are investing heavily in space programs, launching their own satellites, space probes, and human spaceflight missions.

China, in particular, is making significant strides in space exploration. Its space agency, CNSA, has launched several crewed missions, established its own space station called Tiangong, and successfully landed a rover on Mars. While China is not currently a partner in the ISS, its accomplishments demonstrate that space exploration is no longer the exclusive domain of a few superpowers. As China continues to expand its space capabilities, there may be opportunities for future collaboration between NASA, Roscosmos, and CNSA.

India's space program, led by ISRO, has also seen remarkable success, particularly in low-cost missions to the Moon and Mars. India's Chandrayaan and Mangalyaan missions have demonstrated the country's growing capability in space exploration, and ISRO's plans for a crewed space mission could position India as a significant player in international space missions.

As more countries develop their space programs, the potential for international collaboration grows. The challenges of space exploration—such as developing new propulsion systems, protecting astronauts from radiation, and finding sustainable ways to live on other planets—are too great for any one country to tackle alone. By working together, nations can share resources, expertise, and technology to overcome these obstacles and achieve common goals.

International Space Missions and the Path to Mars

One of the most ambitious goals in space exploration is sending humans to Mars. This objective will require unprecedented levels of international cooperation, as no single space agency has the resources or expertise to accomplish such a mission alone. NASA's Artemis program, which aims to return humans to the Moon and establish a sustainable presence there, is seen as a stepping stone to Mars. International partnerships, such as those with Roscosmos and ESA, will be crucial in developing the technologies and infrastructure needed for deep space exploration.

In addition to governmental space agencies, private companies will play a significant role in future Mars missions. SpaceX's Starship program, which aims to develop a fully reusable spacecraft capable of carrying humans to Mars, is one of the most ambitious efforts in this field. Collaboration between NASA, Roscosmos, and private companies like SpaceX could accelerate the timeline for reaching Mars and ensure the success of these missions.

International space missions, particularly those aimed at exploring other planets, represent the future of space exploration. By working together, countries can pool their resources, share scientific knowledge, and develop the technologies needed to push humanity further into the cosmos.

The Benefits of International Space Missions for Earth

While the primary focus of international space missions is often on exploration and discovery, these missions also provide significant benefits for life on Earth. Research conducted on the ISS has led to advancements in medicine, agriculture, and environmental monitoring. For example, experiments on the ISS have helped scientists better understand diseases like osteoporosis and muscular atrophy, leading to new treatments for these conditions on Earth.

International space missions also contribute to global scientific collaboration, bringing together scientists from different countries to work on shared problems. This collaboration fosters the exchange of ideas and technology, leading to innovations that can benefit people for the world. The scientific discoveries made through international space missions not only advance our understanding of space but also have tangible benefits for everyday life on Earth. This exchange of knowledge also ensures that space exploration remains a shared pursuit for humanity, rather than a competition between nations.

The Path Forward for Global Space Collaboration

The importance of international collaboration in space exploration cannot be overstated. As humanity sets its sights on more ambitious goals, such as returning to the Moon and exploring Mars, the need for cooperation between countries and space agencies becomes even more critical. NASA and Roscosmos' long-standing partnership serves as a model for how nations can work together to achieve scientific progress, even in times of political tension. With more countries and private companies joining the ranks of space explorers, the future of international space missions looks bright, and the possibilities for discovery are limitless.

As we look ahead to the next decade of space exploration, it is clear that international partnerships will be essential to unlocking the mysteries of the universe. Whether it is through joint missions to the ISS, collaborations on lunar exploration, or the pursuit of crewed missions to Mars, the spirit of cooperation will continue to drive humanity's quest to explore the cosmos.

CHAPTER 3
Life on the ISS

The International Space Station (ISS) is one of the most remarkable scientific and technological achievements in human history. It serves as a living laboratory, where astronauts from various countries come together to conduct groundbreaking research that would be impossible on Earth. The ISS orbits the planet at a speed of roughly 17,500 miles per hour, completing a full orbit every 90 minutes, allowing astronauts to witness 16 sunrises and sunsets every day. While life aboard the ISS might sound like a dream for space enthusiasts, it is also filled with challenges that test the physical and mental stamina of its crew members.

The daily routine of astronauts aboard the ISS, particularly during the Crew-8 mission, was a well-orchestrated balance between scientific experiments, maintenance tasks, physical exercise, and personal downtime. This chapter explores the unique lifestyle of astronauts in a microgravity environment, the experiments they conducted during the Crew-8 mission, and the physical and psychological challenges of living in space for extended periods.

Daily Life in Microgravity: Routines and Challenges

Life on the ISS operates on a carefully crafted schedule. Astronauts follow a routine that resembles a typical Earth-based workday, albeit in an extraordinary environment. Their days begin with a wake-up call and breakfast, followed by a morning planning conference with mission control. The agenda for the day is usually packed with scientific experiments, station maintenance, and personal activities. However, the unique microgravity environment on the ISS means that even mundane tasks such as eating, sleeping, and exercising require special adjustments.

One of the primary challenges of living in microgravity is learning how to navigate and perform tasks without the help of gravity. Astronauts aboard the ISS float around the station and must rely on handrails, foot restraints, and tethered tools to perform everyday tasks. Simple actions, such as drinking water or eating, are done differently. For example, liquids are consumed from specially designed pouches with straws to prevent floating droplets, and food is packaged in vacuum-sealed bags to ensure it doesn't float away.

Sleeping in space is also an unusual experience. Instead of lying down on a bed, astronauts zip themselves into small, sleeping compartments attached to the walls of the station.

These compartments are designed to offer a sense of privacy and protection from noise, light, and other distractions. Since there's no gravity to pull their bodies down, astronauts can sleep in any position, but most prefer to secure themselves in a comfortable floating position using a sleeping bag.

Astronauts work for about 10 hours a day on weekdays and five hours on weekends, with time set aside for meals, exercise, and communication with loved ones back on Earth. The psychological well-being of the astronauts is of paramount importance, as isolation, confinement, and long-distance communication can lead to emotional challenges. Crew-8, like other ISS crews, benefited from access to private video calls with family members and recreational activities, such as watching movies or reading books, to help them cope with these challenges.

Physical Fitness in Space: The Importance of Exercise

One of the most critical aspects of daily life on the ISS is maintaining physical fitness. In the microgravity environment of space, the human body undergoes significant physiological changes. Without the resistance provided by Earth's gravity, muscles begin to atrophy, and bones lose density. To counteract these effects, astronauts aboard the

ISS are required to exercise for at least two hours each day using specialized equipment designed for space.

The exercise regimen aboard the ISS includes three main pieces of equipment: the Treadmill with Vibration Isolation and Stabilization (TVIS), the Cycle Ergometer with Vibration Isolation and Stabilization (CEVIS), and the Advanced Resistive Exercise Device (ARED). The treadmill and cycle ergometer allow astronauts to perform cardiovascular workouts, while the ARED provides resistance training, simulating the effects of lifting weights on Earth.

The importance of exercise in space cannot be overstated. Not only does it help maintain the astronauts' physical health, but it also contributes to their mental well-being by providing a structured, familiar activity in an otherwise unfamiliar environment. The Crew-8 mission adhered to this exercise routine, ensuring that the crew remained in peak physical condition despite the extended time spent in space.

The Scientific Mission of Crew-8

The primary focus of life aboard the ISS is scientific research. The unique microgravity environment allows for experiments that cannot be conducted on Earth, providing invaluable insights into fields such as biology, physics, and materials science. The Crew-8 mission was no exception, with its astronauts participating in a wide range of experiments designed to advance our understanding of both space and Earth.

One of the key areas of research during the Crew-8 mission was the study of human biology in space. Scientists are particularly interested in how the human body adapts to long-term exposure to microgravity. Crew-8 astronauts participated in studies examining muscle atrophy, bone density loss, and changes in the cardiovascular system. These experiments are critical for planning future missions to the Moon and Mars, where astronauts will face similar challenges over extended periods.

In addition to biological research, Crew-8 contributed to the study of materials science in space. Microgravity offers a unique environment for testing the behavior of materials without the interference of gravity. Experiments on the ISS have led to the development of stronger, lighter materials that could have applications on Earth, such as in the construction of buildings, aircraft, and spacecraft.

Environmental monitoring was another important aspect of Crew-8's scientific mission. From their vantage point in orbit, the astronauts were able to observe and record changes in Earth's atmosphere, weather patterns, and environmental events. This data is invaluable for understanding climate change and its effects on the planet.

Medical Research and Health Monitoring

One of the major concerns for long-duration space missions is the health of the astronauts. The absence of gravity can lead to a range of health issues, including muscle and bone loss, changes in vision, and alterations to the immune system. To mitigate these risks, Crew-8 participated in a number of medical research studies aimed at understanding how the human body reacts to space conditions.

One of the key studies during the Crew-8 mission focused on the effects of microgravity on the human cardiovascular system. In space, fluids in the body are redistributed, which can lead to changes in blood pressure, heart rate, and circulation. Astronauts on Crew-8 wore special sensors that monitored their cardiovascular health throughout the mission. This research is critical for ensuring the safety of astronauts on future missions to Mars, where they will face long periods of weightlessness.

In addition to cardiovascular research, Crew-8 astronauts participated in studies examining the effects of space on the immune system. Previous missions have shown that microgravity can weaken the immune system, making astronauts more susceptible to infections. Crew-8's involvement in these studies will help scientists develop strategies to protect the health of astronauts during future long-term missions.

Maintenance and Operations: Keeping the ISS Running

In addition to their scientific duties, astronauts on the ISS are responsible for maintaining the station's systems and ensuring that everything runs smoothly. This includes regular checks of the station's life support systems, such as air and water purification, as well as routine repairs and upgrades to the station's equipment.

One of the major tasks during the Crew-8 mission was the installation of new solar arrays to upgrade the station's power supply. These arrays are part of a long-term effort to extend the life of the ISS by providing additional power for future research and operations. The installation of the solar arrays required careful coordination between the crew on the ISS and mission control on Earth, with astronauts conducting spacewalks to complete the installation.

Spacewalks, or extravehicular activities (EVAs), are some of the most challenging tasks astronauts face. During a spacewalk, astronauts exit the safety of the ISS and work in the vacuum of space, tethered to the station to prevent them from floating away. Crew-8's successful completion of multiple spacewalks highlights the skill and bravery of the astronauts involved, as well as the importance of teamwork between the crew and mission control.

Psychological Well-Being and Coping with Isolation

While the physical challenges of living in space are well-documented, the psychological challenges are just as significant. Astronauts aboard the ISS are isolated from their families and friends, confined to a relatively small space, and must adapt to a daily routine that is dictated by mission requirements. Maintaining psychological well-being is essential for the success of long-duration space missions.

The Crew-8 mission emphasized the importance of mental health, with astronauts participating in regular video calls with family members and engaging in recreational activities such as watching movies, reading, and exercising. In addition, astronauts have access to psychological support from mission control, who monitor their well-being and provide assistance if needed.

One of the unique challenges of living on the ISS is the disruption of the natural day-night cycle. Because the station orbits the Earth every 90 minutes, astronauts experience multiple sunrises and sunsets each day, which can interfere with their circadian rhythms. To counteract this, the ISS is equipped with specialized lighting systems that simulate a natural day-night cycle, helping astronauts maintain a healthy sleep schedule.

The Role of Teamwork and Communication

Life on the ISS requires a high level of teamwork and communication. Astronauts from different countries and backgrounds must work together to complete complex tasks, conduct experiments, and maintain the station's systems. Crew-8 exemplified the importance of collaboration, with its diverse team of astronauts from NASA and Roscosmos working closely to achieve the mission's goals.

Communication with mission control on Earth is another critical aspect of life on the ISS. Astronauts rely on mission control for guidance, support, and troubleshooting, especially during high-stakes operations such as spacewalks. The success of the Crew-8 mission is a testament to the effectiveness of this communication, as the crew and mission control worked together seamlessly to overcome challenges and ensure the mission's success. The astronauts' ability to work together and communicate effectively was key to overcoming the challenges they faced, from the technical complexities of the mission to the emotional toll of living in space for an extended period.

The Future of Life in Space

As humanity continues to explore the possibilities of long-term space missions, life on the ISS offers valuable insights into what future missions to the Moon, Mars, and beyond will look like. The Crew-8 mission, like those before it, has contributed to our understanding of how humans can survive and thrive in space.

Looking ahead, the lessons learned from Crew-8 and other ISS missions will be critical as NASA and its international partners plan for more ambitious missions, including establishing a sustainable presence on the Moon and eventually sending humans to Mars. The experience of living and working in space will also inform the development of new technologies, such as space habitats, life support systems, and propulsion technologies, that will be necessary for these future missions.

life on the ISS during the Crew-8 mission was a unique blend of scientific discovery, technical challenges, and human resilience. The daily routines, physical fitness requirements, scientific experiments, and psychological challenges faced by the astronauts provide valuable lessons for the future of space exploration. As we continue to push the boundaries of human spaceflight, the experiences of astronauts aboard the ISS will serve as a guide for future missions that take us deeper into the cosmos.

CHAPTER 4

The Science Behind Weather Delays

Weather plays a crucial role in space missions, especially during launch, orbit, and re-entry phases. When planning space missions, weather patterns are monitored closely because they can impact the safety and success of a mission. In the case of the Crew-8 mission, weather delays caused by Hurricane Milton demonstrated how unpredictable environmental factors can affect even the most meticulously planned missions.

This chapter will explore the intricate science behind weather delays in space missions. We will look at how weather conditions such as hurricanes, lightning, wind speeds, and sea conditions can pose a danger during launches, in-orbit operations, and splashdowns. Understanding the science behind these weather phenomena helps us appreciate the complexities of space travel and the need for precise coordination between space agencies and meteorological experts.

Weather's Role in Space Missions

The launch, operation, and landing of spacecraft are complex operations that depend on numerous factors, including optimal weather conditions. From high winds at launch sites to ocean conditions for splashdown recovery, weather impacts nearly every aspect of space missions. NASA and other space agencies rely on detailed weather forecasts to plan safe and successful missions. Any deviation from expected conditions can result in delays, rescheduling, or even mission cancellation.

For instance, hurricane season in the Atlantic and Gulf of Mexico often poses a significant risk to missions launching from Florida's Kennedy Space Center. Thunderstorms, lightning, and wind are other variables that are carefully monitored. If weather conditions are deemed unsafe, space missions are delayed to minimize risk to both the astronauts and the spacecraft.

In the case of Crew-8, Hurricane Milton forced NASA and SpaceX to delay the planned undocking and return of the crew from the International Space Station (ISS). These delays were primarily due to rough seas and high winds, which posed a risk to the crew's splashdown and recovery.

This type of delay underscores how terrestrial weather can affect space operations.

Hurricane Monitoring and Its Impact on Space Missions

Hurricanes, such as the one that delayed Crew-8's return, are one of the most significant weather events that can impact space missions. A hurricane's strong winds, heavy rainfall, and turbulent seas can severely disrupt the operations of a spacecraft during its re-entry and splashdown. For missions returning to Earth, like Crew-8, hurricanes can create unsafe landing conditions in designated splashdown zones.

NASA works closely with meteorologists to track the formation, movement, and strength of hurricanes. Using satellite data, radar systems, and high-altitude aircraft, meteorologists are able to monitor hurricanes in real-time, providing mission planners with crucial information on the safest times and locations for splashdowns. This data is analyzed alongside predictions of storm surges, wave heights, and wind speeds, which can vary depending on the storm's intensity and trajectory.

For splashdown missions, like the return of Crew-8, NASA often schedules re-entry near the Gulf of Mexico or the Atlantic coast, areas vulnerable to hurricanes, particularly during hurricane season. If a hurricane is approaching or conditions become too dangerous, the mission is postponed. In the case of Hurricane Milton, the storm's proximity to the splashdown zone created unsafe conditions that necessitated delaying Crew-8's return to Earth.

The ISS itself can also be affected by hurricanes, though not in a direct way. While hurricanes do not pose a physical threat to the station due to its orbit over 200 miles above Earth, they can disrupt communication and data collection during such storms. Hurricanes are massive weather systems, and from their vantage point on the ISS, astronauts can observe these powerful storms from space. This data, combined with ground-based meteorological data, can be used to better understand storm dynamics and improve weather forecasting models.

Understanding the Splashdown: Safe Landing and Recovery

The splashdown phase is one of the most critical aspects of a space mission's re-entry process. Unlike the days of the Space Shuttle program, where spacecraft would land on a runway, today's missions rely on a controlled re-entry and ocean landing to safely bring astronauts back to Earth. For Crew-8, the splashdown location was chosen based on several factors, including proximity to recovery ships, weather conditions, and sea state.

Sea state refers to the general condition of the ocean's surface, specifically the height of the waves and the speed of surface winds. Calm seas are essential for a successful splashdown and recovery. If the waves are too high or the winds too strong, the spacecraft may capsize or sustain damage during landing, putting the astronauts at risk. Additionally, recovery ships must navigate to the landing site quickly and safely to retrieve the crew.

In preparation for a mission's splashdown, NASA and SpaceX work closely with the U.S. Coast Guard and other maritime agencies to ensure that the recovery zone is free of civilian ships and any potential hazards. Once the spacecraft lands, recovery teams are deployed to retrieve the capsule, extract the crew, and transport them to a medical facility for a post-flight health check.

For Crew-8, hurricane-force winds and choppy seas made the planned splashdown off the coast of Florida too dangerous. The decision to delay the mission's return until conditions improved was made after carefully considering all the safety factors involved. Although weather delays can be frustrating, the priority is always the safety of the crew.

Lightning and Wind: Threats to Launch and Landing

While hurricanes are a well-known threat, lightning and wind also play a crucial role in determining whether a space mission can proceed as planned. Lightning is especially dangerous during the launch phase, as it can damage the rocket and disrupt its electrical systems. A lightning strike during launch can cause catastrophic failure, as evidenced by past incidents.

In 1987, a Soviet Soyuz rocket was struck by lightning during launch, which caused a loss of telemetry data. The mission continued, but the incident highlighted the risks that lightning poses to space missions. NASA's stringent launch criteria include guidelines for lightning, with launches being postponed if there is any risk of a lightning strike. These criteria are based on extensive research into how lightning interacts with rockets and spacecraft.

Wind is another critical factor, especially during the launch and landing phases of a mission. High wind speeds can affect the trajectory of the spacecraft and create instability during ascent or descent. Crosswinds, in particular, can be problematic during launch, as they can push the rocket off course, leading to potential failure. Similarly, strong winds can affect a spacecraft's re-entry angle and speed, leading to a dangerous landing. For splashdowns, high winds can create rough seas that pose a danger to both the spacecraft and the recovery teams.

For the Crew-8 mission, wind speeds and sea conditions were closely monitored throughout the return process. The decision to delay the crew's splashdown was based on forecasts of high winds and rough seas in the recovery zone, both of which would have made the splashdown unsafe.

The Role of Meteorologists in Space Mission Planning

Meteorologists play a vital role in every phase of space mission planning, from the initial launch window selection to the final stages of re-entry and splashdown. Their job is to provide accurate and timely weather data that can be used to make critical decisions about the mission's timing and safety.

During a space mission, meteorologists work around the clock, monitoring weather conditions at launch sites, splashdown zones, and along the spacecraft's flight path. They use a variety of tools, including satellite imagery, radar systems, weather balloons, and computer models to predict conditions that may affect the mission. For splashdowns, meteorologists focus on sea state predictions, wind patterns, and storm activity in the recovery area.

The collaboration between meteorologists and mission control is essential for ensuring the safety of the astronauts and the success of the mission. In the case of Crew-8, meteorologists provided regular updates on Hurricane Milton's progress, allowing NASA and SpaceX to make informed decisions about delaying the crew's return. Without accurate weather forecasting, space missions would be far more dangerous and unpredictable.

SpaceX's Role in Weather Monitoring and Decision Making

SpaceX has revolutionized many aspects of spaceflight, and weather monitoring is no exception. The company has developed advanced tools for monitoring weather conditions, including sophisticated algorithms that help predict how weather will affect launches, orbits, and landings. These tools allow SpaceX to make real-time decisions about whether to proceed with a mission or delay it due to weather concerns.

For the Crew-8 mission, SpaceX worked closely with NASA's meteorologists to monitor weather conditions and assess the risks posed by Hurricane Milton. The company's ability to quickly adjust its plans and delay the mission when necessary highlights the importance of flexibility in space mission planning.

SpaceX's success in weather monitoring is part of its broader approach to making space travel safer and more reliable. By investing in cutting-edge technology and working closely with meteorological experts, SpaceX has been able to minimize the risks posed by weather and ensure the safety of its crewed missions.

Lessons Learned from Weather-Related Delays

Weather delays are a fact of life for space missions, but they also provide valuable lessons for the future. The delays faced by Crew-8 underscore the importance of flexibility and adaptability in space mission planning. NASA and SpaceX's ability to adjust their schedules based on weather conditions demonstrates the importance of prioritizing safety over speed.

These delays also highlight the need for continued investment in weather forecasting technology. As climate change increases the frequency and intensity of extreme weather events, including hurricanes, space agencies must be prepared to deal with these challenges. Advanced weather monitoring tools, combined with improved communication between meteorologists and mission control, will be essential for ensuring the safety of future space missions.

Balancing Ambition with Safety

The science behind weather delays in space missions is complex, but it ultimately comes down to one key principle: safety first. Whether it's a hurricane threatening a splashdown zone, lightning endangering a launch, or high winds affecting a spacecraft's trajectory, weather can have a profound impact on the success of space missions is critical to the safety and success of every operation. Weather, in all its unpredictability, demands that space agencies like NASA and private companies like SpaceX remain agile, adjusting their mission timelines and splashdown plans to ensure that no risks are taken with crew safety. The Crew-8 mission, delayed by Hurricane Milton, serves as a reminder of the delicate balance between human ambition and the uncontrollable forces of nature. Through advanced weather monitoring, coordination with meteorological experts, and a commitment to safety, space agencies can mitigate the dangers posed by adverse weather conditions, ensuring the continued success of space exploration in the face of Earth's most powerful natural phenomena.

While delays are often seen as setbacks, they are, in fact, critical to mission success. The Crew-8 mission's handling of weather-related challenges illustrates the importance of making informed, data-driven decisions, even when those decisions require postponements. As technology advances and space exploration grows more ambitious, weather delays will continue to be a part of the equation. By understanding the science behind these delays and leveraging cutting-edge forecasting tools, space agencies will remain prepared for the inevitable challenges posed by Earth's dynamic atmosphere.

Final Thoughts on Weather and Space Missions

The relationship between space missions and Earth's weather is one of careful observation, planning, and adaptation. Whether monitoring a hurricane, avoiding lightning strikes, or dealing with high winds at sea, meteorologists and mission planners work together to keep astronauts safe. The lessons learned from past missions, including Crew-8, inform future endeavors, ensuring that space exploration continues in a safe and controlled manner.

CHAPTER 5

Hurricane Milton and its Impact

Hurricane Milton played a significant role in delaying the return of the Crew-8 astronauts from the International Space Station (ISS). The mission's timeline, which was carefully planned for months, was suddenly disrupted by the arrival of a powerful storm. This chapter will explore how Hurricane Milton affected the mission's schedule, highlighting the challenges of weather forecasting, the safety considerations that led to the delay, and the broader implications for space missions.

Understanding Hurricane Milton

Hurricane Milton was one of the most powerful hurricanes of the season, forming in the Atlantic Ocean and moving towards the Gulf of Mexico with devastating strength. Categorized as a Category 4 hurricane, Milton brought with it winds exceeding 130 mph, heavy rainfall, and dangerous sea conditions. The hurricane's path threatened key splashdown zones along the southeastern coast of the United States, particularly the Gulf of Mexico and the waters off the coast of Florida, where NASA and SpaceX had planned to bring the Crew-8 astronauts back to Earth.

Tracking hurricanes is a complex process that involves satellite data, meteorological modeling, and real-time observations. In the days leading up to the planned splashdown, NASA and SpaceX teams were closely monitoring Hurricane Milton's trajectory to determine whether it would affect the return of the Crew-8 mission. As the hurricane gained strength and approached the targeted splashdown zone, it became clear that the mission would need to be delayed for safety reasons.

The Science of Hurricane Forecasting

Hurricanes are among the most dangerous and unpredictable weather events on Earth. They are characterized by powerful winds, torrential rains, and surging ocean waves. Forecasting hurricanes, like Milton, involves the use of advanced technologies, including weather satellites, ocean buoys, and high-altitude aircraft that collect data on storm intensity, movement, and potential landfall. These forecasts help mission planners determine whether it is safe for spacecraft to return to Earth.

For the Crew-8 mission, meteorologists relied on these forecasting tools to track the progress of Hurricane Milton and predict its impact on the Gulf of Mexico and Atlantic Ocean splashdown zones. The storm's unpredictable behavior made it difficult to plan a safe re-entry window. Initially, mission planners hoped the hurricane would weaken or change course, but as the storm continued to intensify, it became clear that a delay was necessary.

Meteorological data from the National Oceanic and Atmospheric Administration (NOAA) played a crucial role in the decision-making process. NOAA's data showed that the storm's impact on sea state—measured by wave height and wind speed—would make the waters too dangerous for a splashdown and recovery operation. In addition, the hurricane's wide-reaching effects meant that even distant splashdown sites could experience hazardous conditions, further complicating the mission's timing.

The Decision to Delay Crew-8's Return

As Hurricane Milton continued to move towards the splashdown zone, NASA and SpaceX faced a critical decision. Should they attempt the splashdown despite the storm, or should they delay the mission to ensure the crew's safety? Ultimately, safety concerns took priority, and the mission was delayed until the storm had passed and the waters in the recovery area had calmed.

Delaying a space mission is never an easy decision. A delay can have significant logistical and financial implications, as well as psychological effects on the crew, who are prepared for a specific timeline. In the case of Crew-8, the astronauts had already spent months aboard the ISS, and the delay meant extending their mission even further. However, the risks of proceeding with the splashdown during a hurricane were too great. High winds and rough seas could have caused the spacecraft to capsize or sustain damage during re-entry, putting the astronauts in danger.

In addition to the risks posed by rough seas, the recovery teams responsible for retrieving the spacecraft would have faced hazardous conditions. Recovery ships and helicopters are essential for ensuring that astronauts are safely extracted from the spacecraft after splashdown. Hurricane-force winds and high waves would have made it nearly impossible for these recovery teams to operate effectively, further increasing the risk to the crew.

NASA and SpaceX made the decision to delay Crew-8's return after extensive discussions with meteorologists and mission control. The delay was communicated to the public, and the astronauts were informed that they would need to remain on the ISS until conditions improved. This decision, while frustrating for the crew, was essential for ensuring their safety.

The Psychological Impact on the Crew

Delays in space missions can have a significant psychological impact on astronauts, particularly when they are ready to return to Earth after months in space. For the Crew-8 astronauts, the delay caused by Hurricane Milton meant an extended stay on the ISS, adding additional strain to an already challenging mission.

Astronauts undergo extensive psychological training to prepare for the isolation, confinement, and stress of space missions. However, the unpredictability of space travel, combined with the knowledge that a safe return to Earth is not guaranteed, can create additional mental strain. For the Crew-8 astronauts, the delay required them to adjust their expectations and prepare for a longer mission than originally planned.

NASA provides psychological support for astronauts both during and after their missions. This support includes regular communication with family members, access to mental health professionals, and recreational activities aboard the ISS. For Crew-8, maintaining a positive mindset during the delay was essential for ensuring that the astronauts remained focused and mentally resilient.

Mission Control's Role in Managing the Delay

Mission control plays a critical role in managing space missions, particularly during unexpected delays like those caused by Hurricane Milton. For the Crew-8 mission, mission control was responsible for coordinating with meteorologists, monitoring the hurricane's progress, and making decisions about the timing of the astronauts' return. This required constant communication between NASA, SpaceX, and the recovery teams on the ground.

During the delay, mission control worked closely with the crew aboard the ISS to adjust their schedule and ensure that they had everything they needed for an extended stay in space. This included managing the station's resources, such as food, water, and oxygen, and coordinating with the crew to continue scientific experiments and maintenance tasks.

The delay also required mission control to reassess the timeline for other space missions. The ISS is a busy hub for scientific research and space exploration, with multiple missions planned throughout the year. Delaying the return of Crew-8 had a ripple effect on other missions, requiring careful planning and coordination to avoid conflicts and ensure that the station's operations continued smoothly.

The Financial and Logistical Implications of the Delay

Space missions are expensive endeavors, and delays can increase costs significantly. For the Crew-8 mission, the delay caused by Hurricane Milton resulted in additional expenses related to extending the astronauts' stay aboard the ISS, maintaining the spacecraft, and supporting the recovery teams on the ground.

The logistics of extending a space mission are complex. NASA and SpaceX had to ensure that the ISS had enough supplies to support the crew for the extended duration of their mission. This included managing food, water, and oxygen supplies, as well as coordinating with upcoming cargo resupply missions to ensure that the station remained fully stocked.

In addition to the logistical challenges, the delay also had financial implications for SpaceX. The company's reusable Crew Dragon spacecraft, which was scheduled to return to Earth with the Crew-8 astronauts, was delayed in its return, impacting SpaceX's schedule for other missions. The delay also required additional resources to support the recovery teams, who were on standby for several days while waiting for the storm to pass.

Despite the financial and logistical challenges, the decision to delay the mission was the right one. The safety of the crew is always the top priority, and NASA and SpaceX were committed to ensuring that the astronauts returned to Earth safely, even if it meant incurring additional costs.

Lessons Learned from Hurricane Milton's Impact

Hurricane Milton's impact on the Crew-8 mission highlighted the importance of flexibility and adaptability in space mission planning. While space agencies can plan for many contingencies, the unpredictability of weather, particularly hurricanes, requires mission planners to remain agile and prepared to make adjustments when necessary.

One of the key lessons learned from Hurricane Milton is the need for improved weather forecasting and communication tools. While meteorologists were able to track the hurricane's progress and predict its impact on the splashdown zone, the storm's rapid intensification and wide-reaching effects made it difficult to plan a safe return window for the crew. Advances in weather modeling and satellite technology could help improve the accuracy of hurricane forecasts, allowing space agencies to make more informed decisions about mission timing.

Another important lesson is the need for contingency plans in space mission planning. The Crew-8 mission had multiple splashdown sites available, but the widespread impact of Hurricane Milton meant that all of the planned sites were affected. In the future, mission planners may need to consider more diverse splashdown options or alternative landing methods to avoid delays caused by weather.

The Balance Between Safety and Ambition

The Crew-8 mission's delay due to Hurricane Milton serves as a reminder of the delicate balance between human ambition and the natural world. While space exploration is a testament to human ingenuity and determination, it is also subject to the forces of nature. Weather events like hurricanes can disrupt even the most meticulously planned missions, requiring space agencies to remain flexible and prioritize safety above all else.

For the Crew-8 astronauts, the delay was a challenge, but it also highlighted the resilience and adaptability of both the crew and the mission control teams. By delaying the mission and waiting for safe conditions, NASA and SpaceX ensured that the astronauts returned to Earth safely, demonstrating the importance of putting safety first in space exploration.

As space missions become more ambitious, with plans to return to the Moon and eventually send humans to Mars, weather delays like those caused by Hurricane Milton will continue to be a factor in mission planning. By learning from these experiences and investing in improved weather forecasting and contingency planning, space agencies can continue to push the boundaries of exploration### Chapter 5: Hurricane Milton and Its Impact

Hurricane Milton significantly affected the return of Crew-8 from the International Space Station (ISS), causing delays and bringing the importance of weather management in space missions to the forefront. This chapter will provide an in-depth exploration of how Hurricane Milton played a pivotal role in rescheduling the astronauts' return to Earth, examining the effects of unpredictable weather on mission operations and timelines.

0Understanding Hurricane Milton

Hurricane Milton was a powerful Category 4 storm that developed in the Atlantic Ocean and impacted much of the Gulf of Mexico. It grew rapidly in intensity, with sustained winds exceeding 130 miles per hour, making it one of the most significant weather events during the 2024 Atlantic hurricane season. As it approached the southeastern United

States, it became clear that its trajectory would disrupt key areas, including the splashdown zones intended for Crew-8's return.

The science of hurricane formation and tracking relies on a combination of satellite data, high-altitude weather balloons, ocean buoys, and sophisticated modeling systems. These tools allow meteorologists to predict the path and intensity of hurricanes like Milton. Despite these advanced systems, predicting exactly where a hurricane will go, how strong it will become, and when it will make landfall remains challenging. Hurricane Milton followed a particularly erratic path, initially forecasted to weaken, only to rapidly intensify, further complicating the decision-making process for NASA and SpaceX.

The proximity of the storm to the Gulf of Mexico and the eastern seaboard forced NASA and SpaceX to assess the safety of the astronauts' re-entry trajectory and splashdown location. The rough seas, storm surges, and gale-force winds caused by Hurricane Milton created an environment too hazardous for a safe splashdown, prompting a delay in the mission's return.

The Science of Splashdown and Recovery

Unlike earlier spacecraft, which landed on hard surfaces, modern missions often conclude with a splashdown in the ocean. The decision to use an ocean landing is based on several factors, including safety, ease of recovery, and reduced impact on the spacecraft and crew. However, splashdowns are not without risks, particularly when it comes to weather conditions at sea.

A successful splashdown depends on calm seas and manageable wind speeds. High waves can cause the capsule to flip or become difficult to recover, while strong winds can affect the trajectory of the spacecraft during descent. For the Crew-8 mission, Hurricane Milton's impact on the sea state in the Gulf of Mexico rendered the splashdown site too dangerous, with wave heights and wind speeds exceeding safety limits.

NASA and SpaceX use a complex system of weather forecasting and sea-state monitoring to determine the best splashdown times and locations. This system considers wave height, wind speed, and other factors that could influence the recovery operation. In the case of Crew-8, the weather forecasts for all possible splashdown zones indicated hazardous conditions, leading to the decision to postpone the astronauts' return.

The recovery process itself is a finely tuned operation involving a team of highly trained personnel, recovery ships, and helicopters. These teams are responsible for retrieving the spacecraft from the water, extracting the astronauts, and transporting them to a medical facility for post-flight assessments. A delayed recovery due to rough seas or high winds could pose serious risks to both the astronauts and the recovery team.

Hurricane Monitoring and the Role of Meteorologists

Meteorologists play an essential role in space missions, particularly when it comes to planning safe launches and re-entries. The work of meteorologists in spaceflight involves constant monitoring of weather patterns at various stages of the mission. This includes watching for hurricanes, thunderstorms, lightning, and high winds that could affect the spacecraft or the recovery team.

NASA, SpaceX, and the National Oceanic and Atmospheric Administration (NOAA) worked in concert to monitor Hurricane Milton's development. NOAA provided real-time satellite imagery and advanced weather models that showed Milton's projected path, wind speeds, and the anticipated impact on sea conditions. Using this data, meteorologists were able to predict when the conditions would improve enough for a safe splashdown.

One of the key challenges in this case was the unpredictability of Hurricane Milton. The storm underwent rapid intensification, which caught some weather models by surprise. As the storm moved into warmer waters in the Gulf of Mexico, it strengthened far beyond initial expectations, resulting in a heightened risk for both the astronauts and the recovery teams.

Real-time weather updates allowed NASA and SpaceX to make informed decisions, ultimately choosing to delay the mission rather than risk the lives of the astronauts.

The Decision to Delay: Safety Comes First

Space missions operate with the safety of the crew as the highest priority. Every aspect of the mission, from launch to landing, is planned with multiple layers of safety measures. When it became clear that Hurricane Milton posed a serious threat to Crew-8's return, NASA and SpaceX had to make the difficult decision to delay the mission.

Delays in space missions are not uncommon, particularly when it comes to re-entry and splashdown. Weather conditions, mechanical issues, or unforeseen technical problems can all lead to rescheduling. However, delays can be costly and challenging, requiring adjustments in mission control operations, crew schedules, and recovery logistics.

In the case of Crew-8, the decision to delay the splashdown by several days was made after a thorough assessment of the risks posed by Hurricane Milton. The safety of the crew was paramount, and mission control decided that it was better to wait for the storm to pass than to attempt a dangerous splashdown in rough seas.

The Impact of the Delay on Crew-8 Astronauts

For astronauts, extended stays aboard the ISS can be both physically and mentally challenging. While the ISS is well-equipped to support long-duration missions, the crew must adapt to life in space for longer than planned when delays occur. For Crew-8, the delay caused by Hurricane Milton meant spending additional time in space, requiring adjustments to their daily routines and responsibilities.

Astronauts aboard the ISS have a carefully managed schedule, balancing scientific experiments, maintenance tasks, physical exercise, and personal time. The extended mission meant that the crew had to continue their work while waiting for weather conditions to improve. This could include extending ongoing experiments, conducting additional maintenance, and ensuring that the station remained fully operational.

Psychologically, the delay added stress to an already demanding mission. After months in space, astronauts are often eager to return to Earth and reunite with their families. A delay can create frustration and uncertainty, though astronauts are trained to remain calm and focused under pressure. NASA provides psychological support for astronauts, including regular video calls with family members and access to mental health professionals, to help them cope with the challenges of space travel.

Despite the delay, the Crew-8 astronauts continued to perform their duties and remained in good spirits, demonstrating the resilience and professionalism that are hallmarks of space exploration.

Logistical and Financial Implications of the Delay

Delays in space missions have significant logistical and financial implications. Extending a mission means using more resources, including food, water, and oxygen, which must be carefully managed aboard the ISS. The delay also required adjustments to the schedules of other missions, as

NASA and SpaceX had to ensure that there were no conflicts with upcoming launches or other operations.

The financial cost of delays can be substantial. Space missions are expensive, and each additional day in space adds to the overall cost of the mission. NASA and SpaceX had to allocate additional resources to support the extended mission, including maintaining the spacecraft, coordinating recovery teams, and ensuring that the ISS remained fully stocked with supplies.

In addition to the costs associated with the extended mission, there were also potential impacts on SpaceX's future operations. The Crew Dragon spacecraft used for Crew-8's return was delayed in its recovery, which could have affected the timing of other planned missions. However, the flexibility and adaptability of NASA and SpaceX allowed them to manage these challenges effectively.

Learning from Hurricane Milton: Preparing for Future Storms

Hurricane Milton's impact on the Crew-8 mission provided valuable lessons for future space missions. One of the key takeaways was the importance of flexibility in mission planning. While space missions are carefully planned months or even years in advance, weather events like hurricanes can disrupt even the most meticulous schedules. The ability to adapt and adjust plans based on real-time weather data is crucial for ensuring the safety of the crew.

Another lesson learned from Hurricane Milton is the need for advanced weather forecasting tools. Although meteorologists were able to predict the general trajectory and intensity of the storm, its rapid intensification caught many by surprise. Improved hurricane modeling and forecasting technology could help provide more accurate predictions, allowing mission planners to make more informed decisions about timing and splashdown locations.

NASA and SpaceX will likely continue to refine their weather monitoring systems, incorporating the latest advancements in meteorological science to better predict and manage the impact of hurricanes and other extreme weather events on space missions.

Balancing Ambition with Caution

The delay of the Crew-8 mission due to Hurricane Milton highlights the delicate balance between human ambition and the unpredictability of nature. While space exploration pushes the boundaries of what is possible, it is also constrained by the forces of nature, including weather events like hurricanes. The decision to delay the mission was a reminder that, despite our technological advancements, safety must always come first.

The success of the Crew-8 mission, despite the delay, demonstrated the importance of careful planning, flexibility, and collaboration between space agencies, meteorologists, and recovery teams. As space missions become more ambitious, with plans to return to the Moon and send humans to Mars, the lessons learned from Hurricane Milton will inform future mission planning and help ensure the safety of astronauts in an increasingly unpredictable world.

In the end, the decision to delay the mission due to Hurricane Milton was a testament to the resilience and adaptability of both the crew and the mission control teams. By waiting for the storm to pass and prioritizing safety, NASA and SpaceX ensured that astronauts' return was a testament to the diligence and commitment to safety shared by NASA, SpaceX, and the entire mission control team. The experience with Hurricane Milton has strengthened the resolve to continue improving space mission planning, with an emphasis on better weather forecasting and contingency plans for future challenges posed by nature's unpredictability. As humanity's journey into space advances, the balance between ambition and caution will remain essential in ensuring the success and safety of future missions.

CHAPTER 6
Crew-8: The Astronauts and Cosmonaut

The Crew-8 mission is a testament to international cooperation and scientific achievement. The team consisted of four highly skilled individuals: three NASA astronauts and one Russian cosmonaut, all of whom brought unique expertise and extensive experience to the mission. This chapter provides an in-depth look at the backgrounds, training, and contributions of each member of the Crew-8 team, showcasing how their combined efforts ensured the mission's success.

Michael Barratt (NASA): The Veteran Leader

Michael Barratt served as the commander of Crew-8, and his role was crucial in managing the mission's day-to-day operations, including coordinating with the ground teams, overseeing scientific experiments, and ensuring the crew's well-being. A veteran astronaut with extensive experience, Barratt is renowned for his work in space medicine, a field that focuses on the physical effects of space travel on the human body. His expertise in this area proved invaluable, as the ISS continues to serve as a testing ground for long-term space habitation.

Born in Vancouver, Washington, in 1959, Barratt earned a degree in zoology from the University of Washington before going on to study medicine. He earned his medical degree from Northwestern University, and after completing his residency, he became board-certified in internal and aerospace medicine. His journey into space medicine began with his work at NASA's Johnson Space Center, where he contributed to the development of protocols to protect astronaut health during space missions.

Barratt's background in medicine also led him to conduct research on the physiological impacts of long-term spaceflight, such as bone density loss, muscle atrophy, and the effects of microgravity on cardiovascular health. His scientific contributions were vital to Crew-8's mission, as the crew conducted several experiments aimed at understanding how the human body adapts to the space environment. His leadership and experience ensured that the team was well-prepared for the unique challenges of living and working aboard the ISS.

Barratt had already logged more than 200 days in space before embarking on the Crew-8 mission. His previous missions included a long-duration stay on the ISS in 2009, during which he played a key role in assembling parts of the station and conducting various scientific experiments. His vast experience made him the ideal choice to lead the Crew-8 mission, ensuring the safe and successful execution of its objectives.

Jeanette Epps (NASA): The Aerospace Engineer

Jeanette Epps, a mission specialist on Crew-8, brought an extensive background in aerospace engineering and scientific research to the mission. Her role involved overseeing various technical operations on the ISS and contributing to the scientific research conducted aboard the station. Epps' background as an aerospace engineer allowed her to manage the station's complex systems, ensuring that the equipment and experiments were functioning correctly throughout the mission.

Born in Syracuse, New York, in 1970, Epps earned a degree in physics from Le Moyne College before obtaining a Ph.D. in aerospace engineering from the University of Maryland. Her career began with a focus on research, where she worked on testing composite materials and spacecraft structures. Epps then transitioned to a position at the Central Intelligence Agency (CIA), where she spent seven years as a technical intelligence officer.

Epps' journey to becoming an astronaut began in 2009 when she was selected as a NASA astronaut candidate. Her technical skills and scientific expertise made her a valuable asset to the Crew-8 mission. As a mission specialist, she was responsible for conducting scientific experiments, managing the station's technical systems, and assisting with spacewalks when necessary. One of the highlights of Epps' role during the mission was her work on a series of experiments designed to study the behavior of fluids in microgravity, a field that has implications for both scientific research and the development of new space technologies.

Epps' participation in Crew-8 also marked a significant milestone in NASA's efforts to promote diversity and inclusion within its astronaut corps. Her mission set an example for future generations of aspiring astronauts from diverse backgrounds, demonstrating that space exploration is a field where anyone with the necessary skills and determination can make an impact.

Matthew Dominick (NASA): The Test Pilot

Matthew Dominick, another key member of the Crew-8 team, served as a pilot and contributed significantly to the mission's success through his background in flight test engineering and operational leadership. As the mission pilot,

Dominick was responsible for assisting with spacecraft navigation, managing docking operations, and conducting manual control procedures when necessary.

Dominick, born in Wheat Ridge, Colorado, in 1981, began his career as a naval aviator. He earned a degree in electrical engineering from the University of San Diego before attending the U.S. Naval Test Pilot School. His extensive experience in flight test engineering and combat operations made him an ideal candidate for NASA's astronaut program, which he joined in 2017.

Before being selected for the Crew-8 mission, Dominick had accumulated thousands of flight hours in various aircraft, including the F/A-18 Super Hornet. His background in high-stakes operational environments, combined with his technical expertise in avionics and systems engineering, made him well-suited for the challenges of piloting a spacecraft. Dominick's role as pilot involved monitoring the Crew Dragon spacecraft's systems during launch, orbit, and re-entry, as well as overseeing navigation and docking with the ISS.

One of Dominick's key contributions during the mission was his involvement in a series of experiments related to spaceflight operations and autonomous systems. These experiments are designed to improve the safety and efficiency of future space missions, particularly as NASA and its partners continue to develop more advanced spacecraft for long-duration exploration. Dominick's experience as a test pilot allowed him to provide valuable insights into how spacecraft systems function in the challenging environment of space.

In addition to his technical contributions, Dominick's role in the Crew-8 mission exemplifies the importance of having well-rounded astronauts who can handle both the operational and scientific demands of space missions. His leadership and problem-solving skills were critical to the mission's success, particularly during the complex docking and undocking procedures with the ISS.

Alexander Grebenkin (Roscosmos): The Cosmonaut

The inclusion of Alexander Grebenkin, a Russian cosmonaut from Roscosmos, in the Crew-8 mission highlighted the ongoing collaboration between NASA and the Russian space agency, even amid geopolitical challenges. As a mission specialist, Grebenkin played a crucial role in the operation and maintenance of the Russian segments of the ISS, as well as participating in scientific experiments and spacewalks.

Grebenkin was born in St. Petersburg, Russia, in 1985, and began his career in the Russian space program as a flight engineer. He graduated from the prestigious Moscow Aviation Institute, where he earned a degree in aerospace engineering. After joining Roscosmos, he underwent extensive training at the Yuri Gagarin Cosmonaut Training Center, where he developed the skills necessary to work aboard the ISS, including EVA (extravehicular activity) training, spacecraft systems management, and scientific research.

Grebenkin's presence on the Crew-8 mission underscored the importance of international cooperation in space exploration. Despite political tensions between the United States and Russia, NASA and Roscosmos have maintained a strong partnership, particularly when it comes to operating the ISS. This collaboration allows both agencies to share resources, knowledge, and expertise, ensuring that the station remains operational and productive.

During the Crew-8 mission, Grebenkin was responsible for managing the Russian modules of the ISS, conducting maintenance tasks, and assisting with critical operations such as spacewalks. His engineering background was particularly valuable in troubleshooting technical issues that arose during the mission, ensuring that the station's systems continued to function smoothly.

Grebenkin also participated in scientific research, including experiments related to materials science, biology, and physics. One of the key experiments he worked on involved studying the behavior of biological cells in microgravity, with the goal of understanding how spaceflight affects cellular processes such as growth, reproduction, and mutation. This research has implications not only for human health in space but also for the development of new medical treatments on Earth.

The Importance of Teamwork and Collaboration

One of the defining characteristics of the Crew-8 mission was the seamless teamwork and collaboration between the astronauts and cosmonaut. Despite coming from different countries and professional backgrounds, the Crew-8 team worked together as a cohesive unit to accomplish their mission objectives. This spirit of collaboration is essential for the success of long-duration space missions, where astronauts must rely on each other to perform complex tasks, solve problems, and maintain the station's operations.

The Crew-8 team demonstrated the value of diverse skills and perspectives in space exploration. Each member brought unique expertise to the mission, whether it was Barratt's medical knowledge, Epps' engineering skills, Dominick's piloting abilities, or Grebenkin's experience with the Russian segments of the ISS. By working together, they were able to overcome challenges, conduct valuable scientific research, and contribute to the ongoing success of the ISS program.

The Legacy of Crew-8

The Crew-8 mission will be remembered not only for its scientific achievements but also for its contributions to international cooperation and the advancement of human spaceflight. The teamwork, expertise, and resilience demonstrated by the Crew-8 astronauts and cosmonaut set an example for future missions, particularly as NASA and its partners prepare for more ambitious goals such as returning to the Moon and sending humans to Mars.

Each member of the Crew-8 team has left a lasting legacy, contributing to the growing body of knowledge about space exploration and human health in space. Their work aboard the ISS has paved the way for future generations of astronauts, who will build on their achievements as humanity continues its journey into the cosmos.

In conclusion, the Crew-8 mission exemplifies the best of human spaceflight: scientific discovery, international collaboration, and the pursuit of knowledge beyond our cosmos, continues its journey into the cosmos.

The successful return of Crew-8 demonstrated the importance of safety, flexibility, and cooperation in space exploration. By working together, these astronauts and cosmonaut exemplified what it means to push the boundaries of human achievement. Their mission not only advanced scientific research aboard the ISS but also laid the groundwork for future missions to the Moon, Mars, and beyond.

In the coming years, their contributions will continue to inspire new generations of scientists, engineers, and explorers. The story of Crew-8 reflects the enduring spirit of exploration and the importance of international partnerships in achieving humanity's most ambitious goals in space.

CHAPTER 7
SpaceX's Role in Human Spaceflight

SpaceX, founded by Elon Musk in 2002, has been a driving force in revolutionizing human spaceflight. The company has not only redefined space travel technology but also shifted the paradigm of space exploration through its focus on cost reduction, reusability, and innovation. From its humble beginnings to becoming NASA's key commercial partner, SpaceX has made space travel more accessible, efficient, and futuristic. This chapter will delve into SpaceX's journey, its groundbreaking technology, and how it transformed human spaceflight, with particular emphasis on missions like Crew-8.

The Genesis of SpaceX: Challenging the Status Quo

Before SpaceX entered the space industry, human spaceflight was largely monopolized by government agencies like NASA and Roscosmos. However, these agencies faced high operational costs and slow innovation cycles due to bureaucracy and reliance on non-reusable spacecraft. Elon Musk's vision for SpaceX was built on three key goals: reducing the cost of space travel, making space exploration accessible to private entities, and enabling humanity's long-term survival by creating a space-faring civilization.

SpaceX's early years were marked by challenges, with multiple failed launches of its first rocket, Falcon 1. However, perseverance paid off, and in 2008, Falcon 1 successfully reached orbit, marking a turning point in SpaceX's trajectory. This achievement demonstrated the company's ability to develop reliable, cost-effective rockets and set the stage for more ambitious projects.

One of the most revolutionary aspects of SpaceX's approach to spaceflight is its emphasis on reusability. Traditionally, rockets were discarded after a single use, resulting in enormous costs for each mission. SpaceX sought to change that by developing rockets that could be recovered and reused multiple times, a concept that would dramatically reduce the cost per launch.

Falcon 9 and the Revolution of Reusable Rockets

The development of the Falcon 9 rocket was a game-changer in the field of spaceflight. Introduced in 2010, Falcon 9 was designed with a reusable first stage that could be recovered after launch, repaired, and flown again. This innovation marked a major leap in space travel technology, significantly cutting costs and proving that space missions could be made more sustainable.

The ability to land the Falcon 9 booster, either on a drone ship at sea or on land, showcased SpaceX's advanced control and landing systems. This reusability is crucial for making human spaceflight more affordable and accessible, as the expensive components of the rocket no longer need to be discarded after a single flight. The technology behind the Falcon 9's reusability is a combination of advanced software for guidance, control, and landing, as well as robust engineering that allows the rocket to withstand the rigors of multiple launches and recoveries.

Falcon 9's successful reusability also paved the way for frequent, lower-cost launches, including missions for government and private sectors, such as launching satellites, cargo resupply to the International Space Station (ISS), and crewed missions like Crew-8. The efficiency of this rocket has had profound effects on the economics of space travel, democratizing access to space for more players in the commercial, scientific, and defense sectors.

Crew Dragon: Revolutionizing Human Spaceflight

SpaceX's Crew Dragon spacecraft represents the company's crowning achievement in human spaceflight technology.

Crew Dragon was developed under NASA's Commercial Crew Program, which sought to restore the United States' ability to launch astronauts into space after the retirement of the Space Shuttle program in 2011. The Crew Dragon is an evolution of SpaceX's earlier Dragon capsule, which was initially used for cargo missions to the ISS. Crew Dragon was specifically designed to carry humans, with a focus on safety, comfort, and advanced automation.

The design of Crew Dragon includes an advanced life support system, touchscreen controls, and the capacity to carry up to seven astronauts. It is equipped with the latest safety features, such as an abort system that can jettison the spacecraft away from danger in the event of a launch failure. This level of safety is paramount for human spaceflight, and SpaceX worked closely with NASA to ensure that Crew Dragon met the rigorous standards required for crewed missions.

Crew Dragon's automated systems allow it to dock autonomously with the ISS, minimizing the need for manual intervention from the crew. This feature has been crucial in simplifying mission operations and reducing the risk of human error. However, astronauts still have manual control over the spacecraft if necessary, providing a redundant layer of safety.

One of Crew Dragon's most significant achievements was the successful Demo-2 mission in May 2020, which carried NASA astronauts Doug Hurley and Bob Behnken to the ISS. This mission marked the first crewed flight to launch from U.S. soil since the Space Shuttle era and demonstrated SpaceX's capability to safely transport astronauts to and from space. The success of Demo-2 paved the way for subsequent crewed missions, including Crew-8.

The Crew-8 Mission: A Milestone in Commercial Spaceflight

The Crew-8 mission, a collaboration between NASA and SpaceX, is a significant milestone in the history of human spaceflight. This mission was part of NASA's efforts to continue human space exploration, expand scientific research aboard the ISS, and demonstrate the viability of commercial space travel. The Crew-8 mission, like other Crew Dragon flights, utilized the Falcon 9 rocket for launch and the Crew Dragon spacecraft for crew transportation.

Crew-8's significance lies not only in its technical accomplishments but also in what it represents for the future of human spaceflight. The mission showcased the seamless integration of private industry with government space agencies, proving that commercial space companies like SpaceX could take on a leading role in human space travel.

The Crew-8 mission's success also reaffirmed the reliability and safety of the Crew Dragon spacecraft, highlighting SpaceX's advancements in reusable rocket technology. With the Falcon 9 rocket and Crew Dragon spacecraft designed for multiple uses, the cost of launching and returning astronauts has been significantly reduced compared to previous space programs.

During the mission, Crew-8's astronauts conducted numerous scientific experiments aboard the ISS, ranging from biological research to materials science. The ability to routinely send astronauts to the ISS opens up new opportunities for scientific discovery, as experiments conducted in the unique environment of microgravity often yield insights that are impossible to obtain on Earth.

5. Innovations in Safety and Technology

SpaceX's commitment to innovation has extended to the safety of its crewed missions. The development of the Launch Escape System (LES) is a prime example of how SpaceX prioritizes the safety of astronauts. The LES uses Super Draco engines integrated into the Crew Dragon capsule to rapidly propel it away from the rocket in the event of a launch emergency. This system is capable of functioning at any point during the launch, ensuring that astronauts have a reliable means of escaping danger in even the most extreme scenarios.

The LES was successfully tested during a high-altitude abort test, which demonstrated the system's ability to carry the Crew Dragon safely away from the Falcon 9 in the event of a catastrophic failure. This technology is a critical advancement in human spaceflight safety and has set a new standard for future spacecraft.

Additionally, SpaceX has pioneered the use of advanced materials in its spacecraft and rockets. The company's engineers have developed heat shields, high-performance engines, and innovative avionics systems that allow SpaceX to push the boundaries of what is possible in space travel. One notable example is the PICA-X heat shield, which protects the Crew Dragon during its fiery re-entry into Earth's atmosphere. This heat shield can withstand temperatures of up to 3,000 degrees Fahrenheit, ensuring that astronauts are kept safe during the descent.

The use of cutting-edge software has also been a defining feature of SpaceX's approach. Crew Dragon's flight systems are powered by sophisticated algorithms that enable real-time navigation, monitoring, and control of the spacecraft. These systems ensure precision during docking, orbit adjustments, and re-entry maneuvers, significantly reducing the risks associated with spaceflight.

Collaboration with NASA and the Commercial Spaceflight Program

SpaceX's partnership with NASA has been critical to the success of missions like Crew-8. The Commercial Crew Program, established by NASA in 2010, was designed to promote the development of private spacecraft capable of transporting astronauts to the ISS. This initiative marked a shift from traditional government-led space programs to a model that encouraged private industry to take a more active role in space exploration.

Through the Commercial Crew Program, NASA provided funding and technical expertise to SpaceX, helping the company develop Crew Dragon and its associated technologies. In return, NASA gained access to a reliable, cost-effective means of sending astronauts to space, reducing the agency's reliance on Russian Soyuz spacecraft for transportation to the ISS.

SpaceX's success in this partnership has proven that commercial spaceflight is not only viable but also capable of driving innovation at a pace that was previously unimaginable. The collaboration between NASA and SpaceX has also opened the door for other private companies to enter the spaceflight industry, fostering a new era of competition and advancement.

Beyond the Commercial Crew Program, SpaceX has also worked with NASA on other critical projects, including cargo resupply missions to the ISS and the development of technologies for future lunar and Martian exploration. This partnership continues to push the boundaries of human spaceflight and set the stage for more ambitious missions in the future.

SpaceX's Role in the Future of Human Space Exploration

The success of Crew-8 and other Crew Dragon missions is just the beginning for SpaceX. The company has even more ambitious plans for the future of human space exploration, including the development of the Starship spacecraft, which is designed to carry humans to the Moon, Mars, and beyond. Starship represents the next step in SpaceX's vision of creating a multiplanetary civilization, with the goal of enabling large-scale colonization of the solar system. Starship, like the Falcon 9 and Crew Dragon before it, is designed with reusability in mind, with the goal of dramatically reducing the cost of interplanetary travel. It will be capable of carrying up to 100 people or large payloads, making it ideal for long-term missions to other celestial bodies.

The development of Starship is part of SpaceX's broader strategy to make space travel as routine as air travel. By building a fully reusable spacecraft that can be refueled in orbit and return to Earth for multiple missions, SpaceX aims to break down the barriers that currently limit human exploration of space. This approach will be critical for achieving Elon Musk's ultimate vision: establishing a human presence on Mars.

Starship's progress has been rapid, with multiple test flights demonstrating the spacecraft's potential. While there are still challenges to overcome, particularly related to re-entry and landing, SpaceX is confident that Starship will soon be ready for crewed missions. The company is already working with NASA to develop Starship as part of the Artemis program, which aims to return humans to the Moon by 2025.

SpaceX's Impact on Space Industry and Future Collaborations

SpaceX has had a transformative impact on the space industry. Its innovations in rocket reusability, spacecraft design, and cost-effective operations have set a new standard for space exploration. The company's success has inspired a wave of new private space ventures, with companies like Blue Origin, Virgin Galactic, and Rocket Lab following in its footsteps.

In addition to its commercial success, SpaceX has also played a key role in strengthening international partnerships. The Crew-8 mission, like others before it, involved collaboration with international space agencies, including Roscosmos, the European Space Agency (ESA), and Japan's space agency JAXA. These partnerships are essential for advancing scientific research, developing new technologies, and promoting peaceful cooperation in space.

SpaceX's ability to work with multiple stakeholders, including governments, private companies, and international organizations, will be critical for the future of space exploration. As humanity sets its sights on more ambitious goals, such as establishing lunar bases and sending crewed missions to Mars, SpaceX will continue to play a leading role in making these dreams a reality.

In conclusion, SpaceX has revolutionized human spaceflight by combining innovative technology with a bold vision for the future. The success of missions like Crew-8 highlights the company's ability to deliver safe, reliable, and cost-effective space travel, while pushing the boundaries of what is possible. With Starship on the horizon and plans for interplanetary exploration, SpaceX is poised to shape the future of space exploration for generations to come.

CHAPTER 8

The Return to Earth: Challenges and Triumphs

Returning to Earth after months in space is both a technical and emotional experience, filled with complex challenges and remarkable triumphs. For the Crew-8 astronauts, this return was especially significant, as they faced delays caused by Hurricane Milton, extending their mission and testing their resilience in unforeseen ways. This chapter will explore the various facets of the return journey, from the intricacies of re-entry to the emotional rollercoaster that astronauts experience as they leave the International Space Station (ISS) and touch down on Earth once again.

The Complexity of Re-Entry: A Delicate Dance with Physics

Re-entry into Earth's atmosphere is one of the most critical phases of any space mission. It involves a carefully planned trajectory that ensures the spacecraft enters the atmosphere at the correct angle and velocity. Too steep, and the spacecraft could burn up due to friction with the atmosphere;

too shallow, and it could skip off the atmosphere and back into space.

For the Crew-8 mission, the SpaceX Crew Dragon capsule needed to enter the atmosphere at a precise angle to ensure a safe descent. As the spacecraft hurtles towards Earth, it encounters extreme temperatures caused by the rapid deceleration through the atmosphere. The heat shield on the Crew Dragon is designed to withstand temperatures of up to 3,000 degrees Fahrenheit, protecting the astronauts inside from the searing heat of re-entry.

The forces experienced during re-entry are another major challenge. Astronauts are subjected to gravitational forces (G-forces) that can reach up to five times the force of gravity they experience on Earth. These forces put immense strain on the human body, particularly after months in the microgravity environment of space, where astronauts' muscles and bones weaken due to lack of use. As they re-enter Earth's atmosphere, astronauts must brace themselves for the discomfort and pressure of this intense physical experience.

Crew-8, like other ISS crews, underwent extensive training to prepare for the rigors of re-entry. Their training included simulations of G-forces and emergency protocols in case of an off-nominal re-entry trajectory. This preparation ensured that they were well-equipped to handle the physical challenges of returning to Earth, despite the extended delay caused by weather conditions.

The Role of the Heat Shield and Parachute Systems

The heat shield is one of the most vital components of the Crew Dragon spacecraft during re-entry. As the capsule plunges through the atmosphere, it relies on this advanced shield to absorb and dissipate the intense heat generated by friction. The PICA-X heat shield, developed by SpaceX, is composed of ablative material that slowly burns away during re-entry, carrying heat away from the spacecraft.

Once the capsule has passed through the hottest part of re-entry, it slows down significantly, but not enough for a safe landing. This is where the parachute system comes into play. SpaceX's Crew Dragon is equipped with a series of parachutes that deploy in stages to gradually slow the descent. The drogue chutes deploy first, stabilizing the capsule, followed by the main parachutes, which reduce the capsule's speed to a safe landing velocity.

For Crew-8, the parachute deployment was a moment of anticipation, as it marked the final stage of their return journey. The successful deployment of the parachutes ensured that the capsule gently splashed down in the designated recovery zone. Despite the earlier delays caused by Hurricane Milton, the parachute system functioned flawlessly, demonstrating the reliability of SpaceX's engineering.

Emotional Impact of Leaving the ISS

Leaving the International Space Station is an emotional experience for astronauts. After months of living and working in the unique environment of space, astronauts form strong bonds with their crewmates and become accustomed to the rhythms of life aboard the station. The departure from the ISS signifies the end of a chapter, and for many astronauts, it is a bittersweet moment.

For the Crew-8 astronauts, the departure from the ISS was delayed due to the weather conditions created by Hurricane Milton. This delay added an extra layer of emotional complexity, as the crew had to adjust their expectations and prepare for an extended stay in space. The delay also meant additional time away from their families, further heightening the emotional strain.

Despite these challenges, astronauts are trained to manage their emotions in high-stress situations. NASA provides psychological support to astronauts throughout their missions, ensuring they have access to mental health resources and opportunities to connect with loved ones through video calls and other communication channels. For Crew-8, this support was crucial in helping them cope with the extended mission and the uncertainty of when they would return to Earth.

The Challenges of Readjusting to Earth's Gravity

After months in space, astronauts' bodies undergo significant changes due to the microgravity environment. Muscle atrophy, bone density loss, and changes in fluid distribution are some of the most common physiological effects of long-duration spaceflight. Upon returning to Earth, astronauts must undergo a period of readjustment as their bodies re-adapt to gravity.

The Crew-8 astronauts faced the challenge of re-acclimating to Earth's gravity after their extended stay on the ISS. Simple tasks like standing, walking, and even lifting objects can be difficult in the days and weeks following a return from space. NASA's medical teams are well-prepared to assist astronauts in this transition, providing them with physical therapy and rehabilitation programs designed to restore muscle strength and bone density.

For the Crew-8 mission, the extended time in space due to weather delays meant that the astronauts had to be particularly mindful of the effects of re-adapting to gravity. The recovery process includes a combination of exercises, medical assessments, and careful monitoring of their vital signs to ensure that their bodies are adjusting properly.

Splashdown Recovery and Post-Landing Procedures

The moment of splashdown is a triumphant culmination of a space mission. However, the work doesn't end when the capsule hits the water. Immediately after splashdown, recovery teams are deployed to retrieve the spacecraft and ensure the astronauts are safely extracted.

For Crew-8, the recovery teams were stationed near the splashdown zone, ready to respond as soon as the capsule landed. These teams include divers, medical personnel, and spacecraft technicians who work together to secure the capsule and assist the astronauts. Once the capsule is secured, it is lifted onto a recovery ship, where the astronauts are carefully extracted.

The post-landing procedures are designed to ensure that the astronauts receive immediate medical attention. After months in space, their bodies are vulnerable to the sudden change in environment, and medical assessments are crucial for monitoring their health. Crew-8 underwent thorough medical checks to assess their physical condition, including tests for muscle strength, bone density, and cardiovascular health.

NASA and SpaceX have refined the recovery process over multiple missions, ensuring that the astronauts are quickly and efficiently transported from the recovery zone to a medical facility. This post-landing care is critical for ensuring the long-term health of astronauts returning from space.

The Psychological Journey of Returning to Earth

The emotional journey of returning to Earth can be just as challenging as the physical aspects of re-entry. Astronauts must transition from the isolated, close-knit environment of the ISS to the hustle and bustle of life on Earth. For Crew-8, this transition was complicated by the extended time in space due to weather delays, which may have heightened their feelings of anticipation and anxiety.

NASA's psychological support team plays a vital role in helping astronauts manage the emotional complexities of returning to Earth. This support includes debriefing sessions, mental health evaluations, and continued access to psychological resources. Astronauts are encouraged to speak openly about their experiences, both the triumphs and the challenges, to process the emotional impact of their mission.

For many astronauts, returning to Earth is an emotional rollercoaster. On one hand, there is the joy of reuniting with family and friends, experiencing the natural beauty of Earth, and savoring simple pleasures like fresh food and open spaces. On the other hand, there is often a sense of loss as they leave behind the extraordinary experience of living in space. The sense of camaraderie and purpose that comes with working on the ISS can be difficult to replicate on Earth, and astronauts may experience a period of adjustment as they reintegrate into everyday life.

Overcoming Weather Delays: Resilience in Spaceflight

The delays caused by Hurricane Milton presented a unique challenge for the Crew-8 mission. Originally scheduled to return to Earth at a specific time, the crew had to extend their stay on the ISS as they awaited favorable weather conditions for a safe splashdown. This delay tested the resilience of both the crew and the mission control teams on Earth.

One of the key lessons from this experience is the importance of flexibility and adaptability in spaceflight. Despite the extensive planning that goes into every mission, unpredictable factors like weather can disrupt even the most carefully laid plans. For Crew-8, the ability to remain focused and patient during the delay was a testament to their training and professionalism.

The mission control teams at NASA and SpaceX also demonstrated remarkable resilience in managing the delays. By closely monitoring the weather conditions and coordinating with meteorologists, they were able to identify the safest window for splashdown and ensure that the crew returned to Earth without incident. This collaboration between the ground teams and the astronauts was crucial for the success of the mission.

The Triumph of Safe Return: Celebrating Crew-8's Accomplishments

Despite the challenges of the delay, the Crew-8 mission was ultimately a triumph of human ingenuity, teamwork, and perseverance. The successful return of the astronauts to Earth marked the completion of another chapter in the ongoing collaboration between NASA and SpaceX, showcasing the advancements in human spaceflight technology and the resilience of the human spirit.

The accomplishments of Crew-8 extend beyond the successful splashdown. During their time on the ISS, the astronauts conducted a wide range of scientific experiments that have the potential to advance our understanding of space and improve life on Earth. From the lessons learned from space operations to biological research, the contributions of the Crew-8 astronauts will continue to resonate within the scientific community for years to come.

The Crew-8 mission, despite its weather-related delays, demonstrated the resilience, dedication, and professionalism of the astronauts and ground teams. Their successful return to Earth was a culmination of rigorous training, cutting-edge technology, and a commitment to ensuring the safety and success of human spaceflight. As humanity continues its exploration of space, missions like Crew-8 serve as both a reminder of the challenges we face and a celebration of our capacity to overcome them.

Looking ahead, the lessons learned from Crew-8 will inform future missions, including those aimed at deeper space exploration. The teamwork, technological advancements, and perseverance demonstrated during this mission underscore the ongoing importance of international collaboration and innovation in human spaceflight. As the astronauts of Crew-8 rejoin life on Earth, their accomplishments will inspire the next generation of explorers, scientists, and engineers, driving humanity's quest to explore the cosmos.

CHAPTER 9

Future of Space Travel: What's Next After Crew-8?

The Crew-8 mission, a milestone in human space exploration, is not an endpoint but a crucial stepping stone towards the future of space travel. With each mission, NASA, SpaceX, and other global space agencies are redefining what is possible, paving the way for deeper space exploration, commercial space travel, and even the long-term goal of establishing human presence on other celestial bodies like the Moon and Mars. This chapter explores the implications of Crew-8 for future space exploration and highlights the upcoming missions and technological advancements that will shape humanity's journey into space.

1. The Expanding Role of Private Industry in Space Exploration

Crew-8, like previous missions under NASA's Commercial Crew Program, highlighted the critical role of private industry in human spaceflight. SpaceX has proven that private companies can successfully collaborate with government space agencies to achieve remarkable feats, such as transporting astronauts to and from the International Space Station (ISS). The success of Crew-8 further reinforces the notion that space travel is no longer exclusively the domain of government agencies. Instead, it is becoming increasingly democratized, with private industry leading innovations and lowering the costs associated with space missions.

Looking ahead, companies like SpaceX, Blue Origin, and Boeing will continue to play an instrumental role in advancing human space exploration. SpaceX's reusable rocket technology has already significantly reduced the costs of launching spacecraft, and future developments are expected to push these savings even further. The growing partnership between private companies and space agencies will open up new opportunities, including the expansion of commercial spaceflight, space tourism, and the potential for private research stations in orbit.

In the post-Crew-8 era, NASA and SpaceX are focused on further refining their collaborative efforts. NASA's Artemis program, which aims to return humans to the Moon by the mid-2020s, will rely heavily on private industry to develop and deploy the necessary spacecraft and infrastructure. The Artemis program is a key stepping stone toward more ambitious goals, such as crewed missions to Mars. SpaceX's Starship program, which is being designed to carry humans on long-duration missions to the Moon, Mars, and beyond, is a major part of this future.

The Artemis Program: Preparing for a Return to the Moon

The Artemis program represents NASA's bold vision to return humans to the Moon for the first time since the Apollo missions of the 1960s and 1970s. However, unlike the short visits of the Apollo era, Artemis aims to establish a sustainable presence on the lunar surface, with the long-term goal of using the Moon as a stepping stone for missions to Mars. The success of missions like Crew-8 has helped lay the groundwork for Artemis by demonstrating the reliability of commercial partnerships and the effectiveness of reusable spacecraft like SpaceX's Crew Dragon.

Under the Artemis program, NASA plans to build a lunar Gateway—a space station in orbit around the Moon that will serve as a hub for exploration and a staging point for missions to the lunar surface. The Gateway will provide astronauts with a place to live and work while conducting lunar exploration and will also act as a waystation for future Mars missions.

The Artemis missions will rely on the Space Launch System (SLS), NASA's next-generation rocket, which is designed to carry heavy payloads and astronauts into deep space. The Orion spacecraft, which will transport astronauts to the Gateway and the lunar surface, is currently being developed to work alongside private-sector technologies such as SpaceX's Starship. These spacecraft will enable NASA to establish a permanent lunar base by the end of the decade, providing critical experience in living and working on another celestial body—an experience that will be essential for future Mars exploration.

SpaceX's Starship: A New Era of Space Transportation

As SpaceX continues to build on the success of its Crew Dragon and Falcon 9 rockets, the company is now focused on its most ambitious project yet: Starship. Starship is a fully reusable spacecraft designed to carry large crews and cargo on long-duration missions to destinations like the Moon, Mars, and even beyond our solar system. Unlike the relatively small Crew Dragon capsule, Starship is being designed to support long-term habitation, with the capacity to carry up to 100 people or large amounts of cargo.

Starship will play a central role in NASA's Artemis program, with plans for the spacecraft to deliver astronauts and cargo to the lunar surface. This partnership between NASA and SpaceX represents the next phase of human space exploration, where private industry and government agencies work together to achieve unprecedented goals.

In addition to its role in lunar exploration, Starship is designed to be the primary vehicle for SpaceX's Mars colonization plans. Elon Musk has long envisioned a future where humanity becomes a multiplanetary species, and Starship is the key to making that dream a reality. With its reusable design and the ability to refuel in space, Starship could make interplanetary travel far more affordable and accessible than ever before.

The development of Starship also has implications for Earth-based space travel. SpaceX envisions a future where Starship can be used for point-to-point travel on Earth, dramatically reducing travel times between distant locations. For example, a flight from New York to Tokyo could be completed in under an hour using a suborbital Starship flight. This could revolutionize the way we think about transportation and further blur the lines between space travel and everyday life.

Mars: The Ultimate Destination

While the Moon represents an important milestone for human space exploration, Mars is the ultimate goal. Crew-8 and other missions to the ISS are laying the foundation for the technical expertise and operational experience needed to eventually send humans to Mars. NASA and SpaceX are both committed to making human Mars exploration a reality, and the success of missions like Crew-8 demonstrates that we are getting closer to achieving this ambitious goal.

One of the biggest challenges of a mission to Mars is the sheer distance involved. A round-trip mission to Mars could take up to three years, requiring astronauts to live and work in space for extended periods. The long-duration missions aboard the ISS, such as those completed by Crew-8, are providing valuable data on how the human body responds to long-term spaceflight, including the effects of microgravity on bone density, muscle strength, and mental health. This research will be critical for ensuring the safety and well-being of astronauts on future missions to Mars.

In addition to the physical challenges, there are significant technical hurdles to overcome before we can send humans to Mars. These include developing spacecraft capable of carrying astronauts and supplies over such long distances, as well as creating life support systems that can sustain a crew for the duration of the mission. SpaceX's Starship is being designed with these challenges in mind, and NASA's research on the ISS is helping to develop the technologies needed for long-term space habitation.

Mars missions will also require advancements in propulsion technology to reduce travel times and increase efficiency. NASA is currently exploring nuclear propulsion systems, which could significantly shorten the journey to Mars, making it more feasible for human exploration. These technologies, combined with the lessons learned from missions like Crew-8, will bring us closer to realizing the dream of landing humans on the Red Planet.

The Role of Artificial Intelligence and Robotics in Future Missions

As space missions become more complex and ambitious, the role of artificial intelligence (AI) and robotics will become increasingly important. AI and robotics are already playing a critical role in the operation of spacecraft, the management of space stations, and the exploration of distant planets. In future missions, these technologies will be essential for managing long-duration missions, particularly those that take astronauts beyond the immediate reach of mission control.

AI-powered systems will be used to monitor and maintain spacecraft systems, manage life support systems, and assist astronauts with scientific experiments and other tasks. Autonomous robots could also be deployed on the surface of the Moon or Mars to conduct preliminary exploration, build habitats, and prepare landing sites for future crewed missions.

The use of AI and robotics in space exploration will reduce the workload on astronauts and help ensure the success of missions even in the event of unexpected challenges. Space agencies and private companies are investing heavily in the development of AI systems that can operate in the harsh environment of space, and missions like Crew-8 are providing valuable data on how these technologies can be integrated into future missions.

International Collaboration and the Future of Space Exploration

The success of Crew-8 is a testament to the power of international collaboration in space exploration. NASA, SpaceX, and Roscosmos worked together to ensure the success of the mission, demonstrating that despite geopolitical challenges, space exploration remains a unifying endeavor. The ISS itself is a symbol of international cooperation, with astronauts from different countries working side by side to advance our understanding of space.

Looking ahead, international collaboration will be even more critical as we prepare for more ambitious missions to the Moon, Mars, and beyond. The Artemis program, for example, involves partnerships with space agencies from around the world, including the European Space Agency (ESA), the Japan Aerospace Exploration Agency (JAXA), and the Canadian Space Agency (CSA). These partnerships will be essential for pooling resources, sharing knowledge, and advancing human space exploration.

As private companies like SpaceX continue to develop new technologies, there will be even more opportunities for collaboration between government agencies, private industry, and international partners. The future of space exploration will be a truly global effort, with countries and companies from around the world working together to achieve humanity's most ambitious goals.

Space Tourism: The Next Frontier

One of the most exciting developments in space exploration is the rise of space tourism. While Crew-8's mission was focused on scientific research and space station operations, the technology developed by SpaceX and other private companies is laying the groundwork for commercial space travel. Space tourism companies like Blue Origin and Virgin Galactic are already offering suborbital flights for paying customers, and SpaceX is working on more ambitious plans for orbital space tourism.

Space tourism has the potential to space tourism has the potential to become a thriving industry in the coming years. While Crew-8's mission was focused on scientific research and space station operations, the technology developed by SpaceX and other private companies is laying the groundwork for commercial space travel. Space tourism companies like Blue Origin and Virgin Galactic are already offering suborbital flights for paying customers, and SpaceX is working on more ambitious plans for orbital space tourism.

CONCLUSION
NASA'S CREW 8

The conclusion of *The Story of NASA's Crew 8* marks not just the end of a mission, but the continuation of humanity's efforts to push the boundaries of space exploration. Crew-8 represents a monumental step forward in international collaboration, technological innovation, and our ability to live and work in space for extended periods. The mission has demonstrated that even with delays and challenges—like the weather disruptions Crew-8 faced—the perseverance and resilience of astronauts, engineers, and scientists can overcome obstacles.

This mission stands as a powerful example of how far we've come since the early days of space travel. Decades ago, space exploration was limited to short-duration missions with much uncertainty. Today, thanks to missions like Crew-8, living in space for months has become routine, and the ISS has evolved into a world-class laboratory that benefits life on Earth as much as it furthers our exploration goals. Crew-8's accomplishments, including groundbreaking scientific experiments and flawless operations aboard the ISS, have added to a long lineage of space achievements that are laying the foundation for even more ambitious projects.

In considering the future, Crew-8 will be remembered as an integral part of NASA's roadmap to the Moon, Mars, and beyond. As space agencies worldwide collaborate more than ever, and private companies like SpaceX provide the technology to make missions more efficient, the lessons learned from Crew-8 will inform the next steps in space exploration. This mission showed that technological advancements like the Crew Dragon spacecraft and reusable rockets have made human spaceflight more accessible and sustainable. These advancements are not only key to maintaining our presence in low Earth orbit but also crucial for extending humanity's reach to the Moon, Mars, and possibly other destinations.

The personal stories of the Crew-8 astronauts are a reminder of the human element in all space endeavors. Despite the rigorous scientific work they carried out, these astronauts' resilience during extended stays in space, weather delays, and re-entry challenges highlights the need for psychological support and robust training programs. Crew-8's astronauts faced the physical and mental toll of space travel with unwavering determination, making them role models for future generations of space explorers. The mission's conclusion marks their return to Earth, but their contributions will continue to shape the trajectory of human space exploration.

The lessons from this mission go beyond the technical achievements and scientific discoveries. Crew-8 demonstrated the importance of adaptability in space travel, where unpredictability, such as weather events or technological glitches, can alter mission plans. The delay caused by Hurricane Milton and the eventual safe return of the crew underscore the critical need for patience and flexibility in the face of uncontrollable factors. These lessons are not only applicable to future space missions but also serve as reminders that exploration, in any form, requires resilience and adaptability.

In terms of scientific impact, the Crew-8 mission made significant contributions. Experiments conducted during the mission spanned a range of fields—from biology and physics to material science and medicine—many of which could lead to breakthroughs in understanding how humans can live and thrive in space for extended periods. These discoveries are key to planning longer missions, such as those to Mars, where astronauts will need to be self-sufficient for years. The knowledge gained from Crew-8's research will help future astronauts tackle the physiological and psychological challenges of deep space travel.

As Crew-8 returns to Earth, their mission is a testament to what humanity can achieve when we work together. International cooperation has always been at the heart of the ISS program, and this mission further strengthened those bonds. The involvement of astronauts and cosmonauts from different nations showed that space exploration is a shared endeavor, transcending national borders and political differences. Moving forward, this spirit of collaboration will be essential for tackling the immense challenges that await us in deeper space.

Looking ahead, the story of Crew-8 is a chapter in a much larger book about the future of human space exploration. Missions like Crew-8 pave the way for the Artemis program's return to the Moon, where astronauts will begin living and working on the lunar surface. This, in turn, will prepare humanity for the next giant leap: sending astronauts to Mars. Each mission, including Crew-8, adds to the body of knowledge necessary to make these ambitious goals a reality. The successful return of Crew-8 demonstrates that we are ready for these challenges, equipped with the technology, expertise, and international cooperation necessary to take the next steps into the cosmos.

In conclusion, *The Story of NASA's Crew 8* is not just a mission log but a testament to human perseverance, scientific progress, and the unquenchable curiosity that drives us to explore the unknown. Crew-8's mission will be remembered not only for its immediate achievements but also for its lasting impact on space exploration. As we look to the future, missions like Crew-8 will continue to inspire new generations of scientists, engineers, and explorers who will take humanity even further into the depths of space.